Lambs in Winter

OTHER BOOKS FROM BRIGHT LEAF

LAMBS IN WINTER

Sketches of a Vermont Life
through Seasons of Change

ALEXIS LATHEM

BRIGHT LEAF
BOOKS THAT ILLUMINATE
Amherst and Boston
An imprint of University of Massachusetts Press

Lambs in Winter has been supported by the Regional Books Fund, established by donors in 2019 to support the University of Massachusetts Press's Bright Leaf imprint.

Bright Leaf, an imprint of the University of Massachusetts Press, publishes accessible and entertaining books about New England. Highlighting the history, culture, diversity, and environment of the region, Bright Leaf offers readers the tools and inspiration to explore its landmarks and traditions, famous personalities, and distinctive flora and fauna.

ISBN 978-1-62534-901-9 (paper); 902-6 (hardcover)

Designed by Deste Relyea
Set in Adobe Jenson Pro and Filosofia

Cover design by adam b. bohannon
Cover photo by Tony, *Sheep in Winter*. AdobeStock.com #374286804.

Library of Congress Cataloging-in-Publication Data
A catalog record for this book is available from the Library of Congress.

British Library Cataloguing-in-Publication Data
A catalog record for this book is available from the British Library.

Grateful acknowledgment is made to the following publishers for permission to reprint excerpts from previously published material:

"Off the Trail" from *No Nature* by Gary Snyder, copyright © 1992 by Gary Snyder. Used by permission of Pantheon Books, an imprint of the Knopf Doubleday Publishing Group, a division of Penguin Random House LLC. All rights reserved.

"Spinazzola: Quella Cantina La" from *Selected Poems* by Richard Hugo, copyright © 1979 by Richard Hugo. Used by permission of W.W. Norton & Co. All rights reserved.

"The Moment" from *Morning in the Burned House* by Margaret Atwood, copyright © 1995. Used by permission of HarperCollins. All rights reserved.

"They Are the Last," from *Why Look at Animals?*, by John Berger, copyright © 2009 by John Berger. Used by permission of John Berger estate. All rights reserved.

"The Evening Is Tranquil, and Dawn Is a Thousand Miles Away," from *Oblivion Banjo: The Poetry of Charles Wright*, by Charles Wright, copyright © 2019 by Charles Wright. Used by permission of Farrar, Straus, and Giroux. All rights reserved.

The authorized representative in the EU for product safety and compliance is Mare-Nostrum Group.
Email: gpsr@mare-nostrum.co.uk
Physical address: Mare-Nostrum Group B.V., Mauritskade 21D,
1091 GC Amsterdam, The Netherlands

For Art
and for the ecological farmers,
who are our best hope

CONTENTS

CONTENTS

CONTENTS

PREFACE

For almost two decades, living on a three-acre homestead in Vermont's most populous county, my husband and I have cultivated a large vegetable garden, planted fruit trees and berries, and raised a small flock of sheep, bees, and laying hens. Once a part of a much larger dairy farm, our farm has been a working farm for most of its entire hundred-year history, and we are committed to keeping the land and our two historic barns in agricultural use, without the use of pesticides, and reliant, as best we can, on current sunlight and muscle power.

Our story is one of two city dwellers who came to farming late in life. We do not pretend to self-sufficiency, but we know enough about raising our own food to have an appreciation for what it would take, and to do so if we needed to. The anthropologist David Graeber calls this "play farming," which is by his account what humans practiced for three thousand years before they took up "serious" farming. That is, they complemented farming with other subsistence activities. We do this on a very small scale as an exercise in freedom and personal sovereignty, and as a form of inquiry into what it will take for human and nonhuman life to endure.

The stories and essays in this book, which cover the years I have lived here, reflect the many components of a whole integrated system of living and farming, cooking and eating, working and playing, with each chapter forming a link in an integrated human ecology. We are, in effect, living inside the concentric rings of a permaculture pattern, with our home and kitchen garden at the center, our fruit trees and pastures surrounding it, more rings of annual crops cultivated by an adjacent educational farm, more pasture, and then woods, which become increasingly wild the deeper you go. The farming stories in *Lambs in Winter* are interlaced with accounts of our own excursions into a greater wildness around us, as well as visitations by wild creatures who have penetrated the porous boundary between the wild and domestic: the hawk who hunts our barn pigeons, the many birds who come to nest and forage in our trees and shrubs, the coyotes who know to leave our animals alone because, in return, we bring them gifts.

These pieces are *essays* in the original sense the word, *essai*, in French, meaning *attempt* or *trial*, as they recount our experiments with "natural" farming and our attempts to live without causing harm, however indirectly. *Natural farming* is of course an oxymoron, reflecting a desire to bring together two contraries, the yin and the yang of too much control and too little, the domestic and the wild.

Between the memorable weather events that bracket these pages—the Valentine's Day blizzard of 2007, and the Great Floods of 2023 and 2024, the story that unfolds is unavoidably a story of climate change. For those of us who have chosen to raise animals and to grow our own food, and who find our bliss in the outdoors, weather is what shapes our world. Weather is the ultimate wilderness that we cannot control, and as the weather gets more and more extreme and unpredictable, our efforts at domestication become both more necessary and more precarious. Though I offer no prescriptions, I do hope that these *essais* will yield some truths.

I came to this place from a background of environmental and food justice activism, after years of defending forests and wild rivers in solidarity with forest-based and Indigenous communities. My life and work here are an extension of an ongoing search for an ethical way to live in a time of climate disruption and unprecedented and growing global inequality. It is a life born out of some deep need to connect not only with the food that sustains us but also with our own deep past as human beings. To live in close relation with other creatures, to practice a measure of food sovereignty, has for us been the basis of an engaged life and not an isolationist one. The ideas expressed by E. F. Schumacher in *Small Is Beautiful*, very much alive when I was a college student, have been largely forgotten—that "small is free, efficient, creative, enjoyable, enduring." At a time when there is much cause to despair over democracies in crisis, and the political failure to find solutions to looming environmental crises, perhaps it is time to resurrect them.

Lambs in Winter

Initiation

(The Early Years)

Initiation

Out of a Whirlwind, a Lamb

Valentine's Day 2007. It had already been snowing for two days before the storm unleashed its heaviest load, amounting to three feet of new snow piled onto the snow that had already fallen. I woke up to the white feathering on the branches of trees, and the ground covered in fresh powder. The world had grown quiet, and the movement of headlights along the interstate in the distance had slowed to a dreamy meander. The snow continued to fall, billowing in lacey curtains enclosing our lives, until all the world had become this little stage. The sheep were hunkered down in the barn, the chickens with them, bedding down on the ewes' woolly backs as they like to do in winter.

Our first lambs had been born the week before, and more were expected, which meant we needed to make frequent trips out to the barn. One of a pair of twins was a bottle lamb—my first, as this was my first winter on the farm and my first experience with lambing—who slept her first three nights inside our house, in a bed we made with a hay pillow inside a recycling bin, a perfect size for a newborn lamb. By the third night she was bounding out of her box, trotting and stumbling around the house. In the daytime we left her out in the barn to be with her family, but at night we brought her inside so we could feed her during the night.

By the third day of the storm, the snow was coming down in whirling dervishes, howling and spinning earth and sky into a single restless organism, aquatic and celestial, glacial, pre-uterine. Limbs ripped from trees, the house knocked as if some creature were coming alive inside of it. A whistling. My partner Art shoveled a trough from the back of the house to the barn but it kept filling up with snow, and he kept having to shovel it out anew. The wind reconfigured the snowdrifts, like an artist erasing and redrawing her strokes. By afternoon the woodpile was completely covered, so was our car, and the paddock fence had disappeared beneath the snow. The ribbon of lights from the road went completely dark as we were by then cocooned inside a whiteout.

Art came back to the house from out of the maelstrom to say two more lambs were born, and so I bundled up and followed him back through the hip-deep snow—already the trough had filled again—to find the newborns, one white, one black, the mother ewe standing over the white one, who was standing up on her wobbly legs, shaking and blinking as if drunk on the newness of the world, the black one lying on the ground, alive but barely. We watched them for a while as the ewe licked her white lamb, sniffed its tail, and then lifted her chin in a gesture of approval. She circled around to check on the other, who was not standing up, sniffed him, and then stepped away. A lamb must be able to stand up on its own in order to nurse, and needs its mother's milk to stay warm. We decided to take the weak lamb inside to warm him by the woodstove and get some milk into him—but first we had to get some milk from the ewe, which meant we had to trudge back through the snow to get a container, and we needed to hold the ewe up against the wall while we reached under her to milk her.

The milk worked its miracles and within an hour the lamb was standing, his ears perked up, which is a sign of vigor in a lamb. Hoping the lamb would now be able to nurse, and that his mother would accept him, we brought him back out to the barn, the little boney bundle tucked under my arm, as the snow continued to fall. We set him down with the ewe and his twin, and left him.

Later that night, I opened the back door and heard, through the howling night, a lamb's heart-piercing cry. "Listen!" I said to

Art. "That's a good sign," he said, but I was not so sure. It sounded to me like a cry for help, and so I bundled up and made my way out to the barn through the deep snow where Art's trough had been obliterated once again, as the wailing grew louder. I found, out in front of the creep, curling up before a wall of snow drifting toward the edge of the barn, a bundle of a lamb, releasing a cry so loud it seemed hardly possible it could come from a thing so small. He was screaming toward the house—toward us—a little black worm around which the storm swirled and swallowed the milky stream pouring down from the night. He was calling for us—*Take care of me!*—in the middle of the biggest snowstorm in more than a hundred years, the last great snowfall I have seen in the decade and a half that has since passed, one last loosened page torn from the last millennium, drifting through the universe in all its wild cold beauty. I scooped up this little snail and its bellowing, and brought him inside, and he became my second bottle lamb, one of many more to come.

Coming Home

Some historians would say that "thinkers" are behind the ideas and mythologies people live by. I think it also goes back to maize, reindeer, squash, sweet potatoes, and rice.
— GARY SNYDER

Our farm is in the Winooski River valley, perched on the lip of an old riverbed, on piles of sediment dumped here when the glacial meltwaters, and the ancient river, receded. The house faces an open floodplain and, rising behind it, a chain of mountains, with the bony summit of Camel's Hump appearing out of the foothills as if in the crook of an arm. A ribbon of headlights hems the base of the mountains where a road follows the river, blinking through the night, and behind us, more mountains, but closer, the hills appearing rounded and soft, like the swells and curves of a body in repose. Along the first ridgeline, a brown-dead swath cuts across the forest canopy, a corridor for the power lines that repeat themselves like an echo over mountains and valleys and coastlines, until they reach the distant tundra plains and canyons of the northlands.

Before I moved here I lived most of my life in cities—in London, England, and Brooklyn, New York, but also, for a time, in Portland, Oregon, and Paris, France. I moved to Vermont around the same time as the man who would become my husband, in the early nineties, though we didn't meet until a decade later. He was raising sheep and chickens here, but had no vegetable garden, no flowerbeds and only one old but productive apple tree. Neither was he processing his sheep's wool into yarn and rovings, tapping maple trees, or keeping bees—all of these we would do together in time. We are learning just how much this three-acre plot of land can give—fiber, food, both foraged and cultivated, beauty, and in the woods and meadows behind us, space for our rambles, as necessary as bread and bone. It gives us a structure for our lives, animal companionship, a relation with the wild creatures who visit our gardens and pastures, our trees, rock wall, and wood pile, a life of infinite inquiry into what it means to live as a human being inside a greater community of life.

As a young child living in London, raised by biophobic parents, my only experience of "nature" was the rare family picnic, that cultural oddity by which the indoors is turned out. In Brooklyn, I lived without access to parks or even trees. There was something in me that knew I didn't belong in the city, though I had never known anything else. When I was fifteen, I saved my babysitting money to buy a bicycle for a six-week American Youth Hostel trip through Maine and the Maritimes. I had to pack all my things into two paniers, with no one in my life who could be of any help, having never owned a bicycle, never ridden any distance, and never slept on the ground or in a tent. I did not even know that a foam pad was a thing, and slept on the hard ground every night after riding all day on a seat that was even harder. I had the most overburdened paniers of anyone on the trip and I was the one who predictably lagged behind.

When I left New York for good—having given up my rent-stabilized apartment, and most of my things, except what I carried on my back, I did not know where I was going. I lived for a time in a *chambre de bonne*, in a houseboat, in a sugar shack and a shepherd's hut, and moved from rental to rental once I came to Vermont, until I met Art and moved in with him here. Both of us came to this life with the innocence of city people (though Art was raised in the suburbs) in search of some vague idea of the pastoral, mine formed out of reading *Anna Karenina*, with images of Levin and his scythe. That idea did not include invasive grasses or late blight or day-old lambs dying in my arms. I suppose my ancestors who left the Old

World also had little idea of what they were emigrating to, but they knew they had to leave, even though they carried no maps of where they were going. It is in the DNA of the settler colonist, to pick up and leave, and to give oneself the license to do so—again and again. But I have lived here now longer than I have lived anywhere. I have taken root here, but like a tree with roots that will hold it in place, the winds still want to take it away. I still feel those winds. But I remain—we remain. In Homer's *Odyssey*, the hero's marriage bed is carved out of a great olive tree, rooted in the center of their world; it is the one thing that remains when everything else has been looted or abandoned. I would still be living a wandering life if not for this olive tree, this partnership we have in caring for this place, in learning from the land and the animals as we go.

My childhood was not a happy one, but I was raised by a mother and in a culture that valued the life of the mind and the imagination. For my first two decades of life my teachers were my books; it would be books, then, who would eventually lead me to discover other kinds of teachers—to learn from trees and grass, caribou and salmon, seeds and compost and domestic animals. A poem, said Robert Frost, begins in delight and ends in wisdom, and so does a garden, and in between the delight and the wisdom are many losses, plagues of flea beetles and varroa mites, the weeding and all the mucking out.

I have lived here long enough to have seen the land change: the trees we planted now fill in what were once open spaces; the old elm is gone, leaving behind a great hole in the sky; the floodplain across the road that was for all those years planted with industrial corn is now a hayfield; and the largest remaining tract of unbroken farmland has been subdivided, the meadows turned into lawn. Mary Jo and Everett Andrews, the elders of the oldest farm family in town, no longer wave to us from their front porch as we pass. We have also seen the educational farm, which set up next door in the newly renovated West Monitor Barn, restore the lands around us and bring this community new life.

Ecological succession is the process through which biological communities evolve over time, passing through the stages of initiation and disturbance before reaching a steady state. Those stages also describe our own experience over two decades of living on and working this land. The early years of a garden are like the halcyon days of the first farmers, when our tomatoes, squashes, and beans spilled from the vines before the arrival of pests and diseases that we have had to manage ever since, before the invasive plants spread over the pastures like spilled ink on the page. We have tempered our early ambitions, given up on bees, after losing two hives in a row; on strawberries, which we could not keep weeded; on blueberries, which the birds always got first; and have come to accept that much of our time must be devoted to cutting back thistle, poison parsley, and knotweed. We would, in time, let our ewes live out their lives but no longer have lambs. But as we settle down to accept our limits, the basis of our lives grows more and more unsettled with a changing climate.

I am reluctant to call myself a "farmer" because we grow food only for ourselves, and the surpluses that we give away, not for a larger community. We are not professionals. "Hobby farmer" isn't right either, as for several years, we sold some lamb at harvest time, yarn and rovings, dried flowers, and, while we were raising bees, a honey-sweetened homemade ice cream, at our local farmer's market, in an earnest effort to reduce our need for wage employment. "Gardener" doesn't fit either, because it suggests someone who tends to roses and peonies, not plants for table and pantry, and besides that we also raise sheep and chickens. I am well aware that the skill and technology applied by the farmers next door to us— who produce some fifty thousand pounds of food in a season—are well beyond those of this home gardener. Sharon Aysk and Aaron Newton, authors of *A Nation of Farmers: Defeating the Food Crisis*, remind us that this is what farmers always did—they didn't just produce one or two crops for market but also kept a cow and some chickens, grew beans and potatoes for their own kitchen. A half a century ago, in our parents' lifetime, almost half of Americans had vegetable gardens, producing much of the food they ate at home. We need a million of these farmers, Aysk says, because *this* is how to feed the world. We cannot count on Monsanto; we will have to do it ourselves, saving our seed and raising animals who will renew themselves, generation after generation. Acting as part-time farmers is to practice what human beings practiced for thousands of years before they invested in "serious" farming, write David Graeber and David Wengrow in *The Dawn of Everything*. Farming was one among a variety of subsistence activities that together gave people flexibility and resilience, in the same way that a diversified farm is more resilient than a monoculture. "The ecology of freedom ('play farming,' in short) ... describes the proclivity of human societies to move (freely) in and out of farming; to farm without fully becoming farmers; raise crops and animals without surrendering too much of one's existence to the logistical rigours of agriculture." *You can't do just one thing* is a premise of ecological farming, as it is a principle in an ecology of freedom, where the wholeness of the human being is restored, and where my freedom, security, and prosperity are dependent upon yours.

And for most of that history, human beings lived in community with one another and with other creatures by participating in a gift economy. To harvest food from a garden, to shear wool from sheep, is to experience life as gifting, and this is no less true because we might sell some of our wool or produce at a farmer's market, nor is it less true because the beans and tomatoes I harvest require a lot of work from me in return. But to experience the proliferation of fruits from the vines, the miraculous abundance, the metamorphosis of nothing but grass, water, and sunlight into the most useful fiber humanity has ever known—I cannot help but feel that I am the recipient of a gift. I might have prepared the soil, weeded, thinned, and weeded again, but this cannot explain it, the gifting nature of soil and sunlight. Gifts occur in "a realm of humility and mystery," Robin Wall Kimmerer writes. This is why the food we grow ourselves is so deeply satisfying. This is why a bowl of beans, as Kristin Kimball writes, and someone to share it with, is enough.

In the early days of living here, we spent so much of our time in the garden and fixing up the house that we found ourselves getting out less and less to ramble in the mountains like we once did. One day I came to miss this—as someone who spent years advocating for wild forests and wild rivers, spending weeks at a time in the back country, sleeping in a tent and cooking over a stick fire. It is true that as human beings domesticated plants and animals, we in turn were domesticated: our range of movement contracted, our senses dulled, and we lost the participatory alertness characteristic of wild creatures. "As our domestic animals settled in for a life of reduced activity and stimulation, so did humans," Carl Safina writes in *Beyond Words*. "As people provided safer, more sedentary conditions for their livestock, they did the same for themselves. The confinement was mutual." I worried that this was what was happening to us, as our world retracted and our engagement with a wider world of wildness grew dimmer. We know that food crops do better in proximity to forests and healthy grasslands, and I believe that it true for us, too, who as domesticated animals thrive better in proximity to wild forests and the nonhuman animals who are not born and bred for our use. We would learn to scale back our ambitions as growers who in our enthusiasm wanted to do it all—beekeeping, fiber arts, agroforestry—who wanted to grow every vegetable that was possible in our climate. Applying the permaculture principle of *self-regulation*, we would need to draw and accept our limits. This is another way of saying we have divided our hours, as Helen and Scott Nearing did, between "bread labor" (what we do to feed and house ourselves, which means for us not only self-provisioning but also wage employment), "personal pursuits" (playing music, writing, going for walks, paddling, skiing, and cycling), and "civic work" (in our case this has meant, mostly, ecological, anti-war, and antiracist activism).

I have never felt that the pleasure or joy I get from flowers, or from the meals we make from the food we grow, are irrelevant to my activism, or to my writing life. George Orwell, near the end of his shortened life, lived on an island in the Hebrides where he grew fruit trees, all kinds of flowers, and vegetables. He shot rabbits, he milked goats, he fished, he cut his own peat for fuel. He had to walk eight miles from the nearest road, which he did often. He was raising his son, on his own, after his wife died. Orwell himself was dying, but before he would die, he would plant lupins, roses, apple trees, cherries. He would write *1984*.

Though we may not think of roses when we think of Orwell, Rebecca Solnit wrote a whole book on Orwell and his roses. She attributes his fidelity to the concrete and the particular, his abhorrence of ideological abstraction, and his commitment to independent thinking to his milking goats by hand and tending roses. In planting his roses in a very particular soil, Orwell is rooting himself in the specific and the concrete, which is the basis of writing well. He would teach generations of writers to use concrete language as a guardrail against the vagueness and insincerity of inflated speech. Solnit also wants us to see that Orwell, while enduring illness, while mourning the death of his young wife, while living through the London Blitz, and while thinking and writing about Stalin's gulags, made room in his life for joy and the pleasures of growing roses. "Our job," Orwell wrote in his essay on Gandhi, "is to make life worth living on this earth, which is the only earth we have."

To say that humans have always lived this way is not to deny the benefits of modern technologies. I, for one, would not want to do without antibiotics, vaccines, or the washing machine, nor without a way of knowing the world that modern science has given us. Our state- of-the-art woodstove is vastly more efficient than that old pot-bellied stove or the open hearth. But to live with more simplicity and closer to the basis of our lives is one way to investigate what it means to be human, and it is to recognize that the benefits of technologies that depend on resource extraction come with substantial penalties. Conditioned as we are by the myth of human progress, we put too much faith in futuristic, unproven technologies when there may be answers right there for the taking in our past—even our recent past—because we have known how to live with less complexity, with less (or no) dependence on fossil energy. For all but a few seconds before midnight in the scale of time, human beings produced their own food. And we—not everyone, but many more of us—will have to do so again.

"Cities are the solution," I heard a panelist say, in a discussion of environmental writing. Learning to green our cities and to humanize them will be necessary as we face a future of ten billion and counting, but the migration from the countryside to the city over the last century has come at the expense of the deliberate devastation of rural places. Cities cannot feed themselves, thus people who live in cities will be forever detached from the problems that come with extracting the resources they depend on and producing the food they eat. There can be no backyard chickens or kitchen gardens— which will be, in my mind, a key part of the solution—in dense urban centers. The countryside cannot be made a sacrifice zone for the city, with industrial feedlots confining animals twenty-four seven, polluting the air and water and killing the life of soils, with migrant laborers held *encerrado* on the farm. The countryside, at least here in the United States, has a car dependency problem it has not even begun to address, but the future of our life on earth will depend on the countryside also being "the solution," as Gandhi saw village life, rather than urban development as the future for nonindustrial nations. I am, at heart, a city person as much as I have learned to be a rural one; cities have made me who I am, a person who chose to live this life—but I will invest my hope in the village and the country as the solution, as the future of wilderness, on the one hand, and cities, on the other, will depend on us taking care of the countryside between.

Lambs in Winter

It was a freezing cold morning in early January when Art woke me with the news:"We have a lamb." In the early years, when we kept a ram with the flock, this was not unusual. It was ten degrees below zero. I could see them from the kitchen window—the all-white ewe standing apart from the others with her all-black lamb standing beside her.

The world outside was covered in ice.

Left alone the ewes will go into estrus on the first cold nights, usually in September but sometimes in late August, which means that if there is a ram in the family the lambs will arrive in the middle of winter.

I went out to find that the ewe was ignoring her lamb. Art had just given them their day's ration of hay. It's always mayhem when the sheep are fed in the morning—they move from hay pile to hay pile, as if someone else's pile might be preferable to one's own, when all of it comes from the same bale.

I picked up the lamb and put my finger in her mouth—cold. She was getting that hunched look—also not a good sign. A lamb needs to nurse to maintain its body heat; if the mouth feels cold, that's a sure sign it's not nursing. I tucked her under my arm and carried her over the paddock fence and across the ice to the house as she wailed. A lamb's cry has evolved to be loud enough to be heard from clear across an open pasture. After a minute or two in front of the woodstove she quieted down and went to sleep.

I let her warm up and then brought her back out to the barn. Art and I tried to get the ewe and her lamb into the lambing pen. A ewe will hear her lamb crying from inside the pen and will usually follow. But this ewe just stood there, half in, half out.

"She needs hay," I said.

Then a different ewe, Ewelysses, barged into the pen and sniffed the lamb as if it were her own. She went out again, then came back, responding to the lamb's cry as only a mother ewe should.

"Wait! That's not her lamb!" Ewelysses was taking a greater interest in the lamb than the lamb's own mother. Ewelysses moved around restlessly. Then she lay down in a far corner of the barn, leaning against the wall.

"She's going into labor," I said. Otherwise, she would not be lying down at mealtime, uninterested in her breakfast.

She got up and ambled into the lambing pen.

She started breathing heavily. She curled her upper lip, making an equine grimace, lifting up her chin. She squirmed and shifted her weight from side to side. She got up, turned in a circle, then lay down again. She began to push, groaning slightly.

I had seen ewes give birth before, and it usually seemed so easy. This time it did not. I looked at Ewelysses's backside—and saw the lamb's head and a forepaw emerge.

The ewe continued to strain.

"I need to see the other foot." I walked around to get a better look. Then Ewelysses stood up. She swung around. The lamb was half emerged now, hanging down between the ewe's hindlegs, sheathed in an ochre-colored mucus. I could see both forelegs and the head. Ewelysses circled around and around, the lamb swinging from her as in a game of airplane. It was odd—as if the ewe were trying to shake the lamb loose. The lamb's head knocked gently against the wall. Then it slipped free, landing on the floor like a wet fish on dry ground.

Ewelysses immediately went to work licking the lamb clean. The lamb's head perked up, its eyes opened, and it gave a little cry.

It was convenient that Ewelysses gave birth right there in the lambing pen. The only problem was, we needed the pen for the other lamb and her mom. That lamb was still not nursing. She wandered around—all black with splashes of white on her face like spilled milk—a little lost lamb. I saw her wander outside and curl up on the ice to take a nap.

It was a forlorn sight. I picked her up and brought her back inside the house. As I took off my boots, the lamb stumbled—her hooves sliding across the wood floor—into the closet by the front door, where she settled down. She looked up at me and cocked her head, her ears flopped over like a puppy's ears.

"Not there," I said, and put her next to the stove.

Meanwhile, we needed a second lambing pen. Art went into town to get pallets and eyelets, and quickly constructed a new pen. He had to drag the ewe on her back into the new enclosure, then righted her onto her side.

A lamb's life is delicate in the beginning. Especially when it is cold. A lamb has only so much time to get up on its feet and to figure it all out. Without milk, it cannot keep warm and will soon go into decline. When it is very cold, that window is perilously short. A newborn will struggle to lift her head, and to manage those long gangly legs with their bulbous knees. She will attempt to swing herself over onto her knees, then to unfold her legs, hindlegs first. She will swivel and sway, toppling over a few times before she will get it right. Still sticky and wet, with the so-recent memory of her steamy hot bath inside the womb, she stands beside the ewe a bit dumbly, as some dim consciousness begins to stir deep inside of her that will tell her to turn toward her mother's teat, hanging from its wooly firmament, as if to true north. The ewe will stand patiently as the lamb nudges in all the wrong places, turning her head around to smell her lamb's little tail—and when the lamb at last discovers the secret of life, when she has mastered the gestures of nudging and pulling and sucking, and she drinks—then that little tail will tremble with happiness. And if there is a shepherd standing over them, she will know that all is well and as it should be.

I put the lamb onto the ewe's nipple, and she drank. She drank and drank and drank.

Ewelysses, meanwhile, continued to lick clean her lamb. I watched as the yellowish bundle of wool and bones tried to stand up—first straightening out his hindlegs, then pitching forward as he unfolded his forelegs and swiveled around, before collapsing back into the straw.

Ewelysses doted on her lamb. She stood up for him to nurse; when he was lying down, she stood over him and nudged him, insisting that he get up and nurse, like a mother coaxing a child to eat his spinach. When I went out to check on them in the evening, I saw the lamb shivering as he slept beside her. Ewelysses moved closer and pressed her body against him. I saw the lamb climb up onto her back to nest on her fluffy wool fleece. That is how they'll sleep together during a cold winter night.

Both of the lambs were doing fine. But we worried. Temperatures were expected to fall to twenty below that night. We had never had new lambs arrive in weather like that. Later, I walked back to the barn in the moonlight, the snow creaking under my boots, and as I approached the barn I could hear the cry of a newborn. I shined a light toward the barn and could see a black lamb writhing on the ground. Another lamb. The ewe stood over it, licking it and grunting. The white one was standing beside them. I watched as the ewe moved back and forth between them, licking and nudging the new one, then turning away to attend to the other who needed to nurse.

The long cold spell did not end until all the ewes had delivered their lambs. By February, we could see them all lounging around in the sunlight in front of the barn. At dusk, they would play together, racing and jumping and climbing onto the backs of the ewes. The world was still an enormous snow bowl that would so marvelously become—what the lambs had yet to discover—a miracle of grass and shade trees in the summer sun.

Off the Trail

It is mud season down here in the valley, when we have to stay off the trails, but higher in the mountains there is still snow. We slog up a Bolton Valley main trail on snowshoes, and from there cut through the trees to a narrow winding trail over a ledge. There the trail takes a narrow hairpin turn before it levels off to follow the ridgeline, where the tortured tops of the trees are silhouetted against a blue sky.

> We are free to find our own way
> Over rocks—through the trees—
> Where there are no trails . . .

Gary Snyder's "Off the Trail" was our wedding poem. Art often wants to bushwhack, to venture further, deeper in to the trees, onto the steeper slopes, but I do this only when I am with him, as I would get lost for sure and I am not so bold. It is true what the poem says, that

> Our trips out of doors
> Through the years have been practice
> For this ramble together . . .

If we'd stuck to familiar trails, we would never have found this place in our lives, raising sheep and playing guitar and writing poetry. I trust that he knows his way around this mountain. We are almost in the alpine zone now. We climb further up, until we are looking over a snow bowl with a view over Bolton Mountain, the shaved bands of white pouring down where the movements of skiers are visible, traversing the slopes. Here where the trees are exposed, coated in ice, the birches branch like white capillaries against the blue sky. Spruces are hung with icicles, finger length. In the wind they rattle and chime like chandeliers rustled by a breeze through an open window. A strong gust cracks the ice fingers from their branches; they rain down with the sound of foil crinkling. The light catches the glass crystals as they fall.

I follow Art back below the tree line, up and over to a ledge where there is a view of Mount Mansfield and its range. We backtrack and pick up the trail again, then cut into the woods toward another ledge. Here there is a view of Lake Champlain. We cross over then face the other direction, toward Mansfield, the peak an ugly notch, like a disfigured elbow. Ground out by glaciers. There is a long view over the range, the valley, and then the range rises gradually, like someone getting up from the floor onto her knees.

We come back down Moose Glen, where there actually are moose tracks, though not recent ones. We climb down to another snow bowl, the mountain rising up from either side like great snowy owl wings, and we climb down a trail that swishes from side to side like milk sloshing in a bowl. We cross a small bridge. I can hear music now coming from the ski lodge. At home the house still smells of maple sap slowly cooking on the woodstove, though the sugaring season is already over, the bottles of syrup capped and stored, enough, we hope, to carry us through another year, another cycle of the seasons.

A Barn, a Nest,
a Mooring

The cob barn was listing like a ship in trouble. It was wide open to the weather, having lost its windows and doors long ago. Still, it was a beautiful structure, with its weathered sheathing, horizontal clapboard on one side and vertical planks, spaced to allow air and light, on the other, like an Agnes Martin painting, and a worn slate roof in its many subtle hues. If only it could be lifted and set right. If only it could be made useful again.

It would be expensive to fix. If you see so many old barns falling to ruin, trembling where they stand as if a single gust would bring them down, if you see them standing today and gone tomorrow, this is why. They no longer serve any economic purpose now that the farms are gone, and even if they're not. No one is drying corn this way anymore.

The cob barn—also called a *corn crib barn*, what we referred to as the *little barn*, and later, once we installed Art's electric baby grand there, the *piano barn*—was one of three historic barns on our property, still standing when Art arrived. Originally the barn was designed for drying and storing corn, what they called *Indian corn*, meaning maize, the kind that was dried on the cob and ground into flour or cornmeal, and would after generations of selective breeding become the ubiquitous No. 2 Corn. Besides the cob barn there's the larger carriage barn, originally used for horses, where the sheep sleep at night and live in the winter, and where the chickens go to roost.

The hay loft is still a hay loft. Only the largest barn, the cathedral-sized West Monitor Barn, is gone—moved down the road and restored to its former glory, the collective achievement of the land trust, community activists and volunteers, state and federal agencies, and youth corps work crews. It stands today as one of Vermont's proudest architectural landmarks. We still get to admire it from our window, and as we come out of the woods from our rambles and look toward home. Left behind is its footprint, a deep gouge in the hill where the high-drive was once banked into the slope.

Even in my time here, I have seen the last of the old barns in town vanish one by one. There was the Andrews' barn down the road that burned one night in 2014 in a fire so hot there was hardly a trace of it left behind, only a char-burned patch of ground. In a rather cruel irony, the fire was sparked by an overworked sump pump in a heavy January rain. Next was the Gothic arch balloon-framed barn near the Jonesville bridge that was dismantled and replaced by a garage. I miss that one, too. Our two barns, part of the monitor barn complex of farm buildings, are among the last of the old relicts to anchor us to our agricultural history, and they stand out between all the spilt-levels and vinyl-sided detached houses with their front lawns and ornamental shrubs.

They stand out among the modern barns as well.

"The old barn is fast disappearing," one commentator wrote in May 1948, "and the nation's farms are in for a facelifting. . . . 'Inefficient,' say today's farm experts, pointing to the time and effort wasted in the wood framed barn."

The historic barn, for all its wasted time and effort, is a beautiful thing because it is *vernacular* architecture, unlike the cookie-cutter buildings made of industrial, mass-produced materials that are the same here as anywhere else. They're designed to meet local needs, made of local materials, built in the "common, everyday language of ordinary people in a particular locality" (*Webster's* definition). That is why they seem to belong, *nestled* into the hills and valleys of the landscape, as the metal and concrete structures built of manufactured and assembled parts do not.

Ours, too, is a modern house, a duplex, built in the seventies after the old farmhouse burned down. Over the years we have done much to rid the place of its seventies feel, ripping out carpets, replacing hollow core doors and ceiling tiles. Art did much of the work himself, with some help from me. He put in windows and floors, repaired the roof, and rebuilt the porch. I'm nervous when he does electrical work or plumbing, but he knows when a job is beyond him and it's time to call a professional. As someone who grew up in a household that did not have a working pair of scissors, I admire his ability to fix things and to build. An autodidact in most everything he does, he didn't learn any of this from his banker father. We could not do this if at least one of us was not handy with a hammer or an axe.

I do love old houses, but ours is not one of them. The Queen Anne style, wood-framed houses with their steep gabled roofs and gingerbread cornices, beside steepled churches and English barns, give New England its iconic look but are far outnumbered by the prefabs, the trailers, the ranch-style houses, and the seventies duplexes like ours. I miss the smell of old wood, the narrow staircases that creak underfoot, the wide floor boards cut from the once massive trees that are but a memory. This describes the Brooklyn brownstone I grew up in, and the big rambling Victorian farmhouse I lived in for two years when I first moved to Vermont. I do not get to live in an old house, but we do have our old barns.

With a matching grant from the Vermont Department of Historic Preservation in 2008, we were able to restore the barn. Art hired Eliot Lothrop (hiring a professional was a condition of the grant), a fine craftsman with experience restoring old barns, someone who would respect its architectural integrity. (Another builder we consulted asked us if we were going to turn the barn into a garage.) He would jack up the whole structure with cribbing, dig a trench around the foundation and fill it with crushed stone, build new sills with scarfed joints and mortise and tenon joinery—using wooden pegs and no nails—and bring the whole structure back down to sit on dry stone piers.

In one day the crew had the barn jacked up onto cribbing, and on the second day the excavation had begun, when the excavator hit something large and circular underground. I came outside to find the whole crew standing in a circle around a giant metal ball, the size of a small car, dangling from the excavator. In time we identified the object as a ship's mooring. I submitted a short article to our town newspaper, hoping that someone would come forward with a story about how a World War II–era ship's mooring got buried beneath a small barn in landlocked Vermont.

> **Ship's Mooring Found Underground**
> Work is underway on the restoration of the historic cob barn at the Herttua property on Route 2 in Richmond. The restoration project is funded by a matching grant from the Vermont Department of Historic Preservation. Restoration work is being done by Building Heritage, Ltd.
> On the second day of construction, a large circular metal object was found buried underground beside the barn. The object has been identified as a ship's mooring. How it got there or who buried it is a mystery.

But no one ever did come forward with any knowledge of this thing.

Maybe, I wondered, it was placed there to anchor the barn, to keep it from listing or blowing away, to tether us to our past.

I am reminded of those wooden houses on the coasts of Newfoundland, a place we're both drawn to, anchored to rock to keep them from blowing away.

If I had dreamed this—a ship's mooring uncovered beneath a barn—I would have found meaning in this: for those who have traveled, the wandering life still tugs at us. For those with a love of deep lakes, wild rivers, and rock-strewn seacoasts. I am reminded of a curious moment at the end of Homer's *Odyssey*, when Odysseus reaches the farthest inland place in the imagination of a sea-faring

Greek, where his oar is mistaken for a winnowing fan. Perhaps our ship's mooring was misunderstood, as was Odysseus's oar; perhaps it was repurposed to anchor a barn, or to store contaminated gasoline (a foul brown liquid did leak from it when it was lifted), or to bury some other secret. Perhaps it has resurfaced to tell us something about ourselves.

In 2023, Lothrop will lift and repair the East Monitor Barn down the road, the largest of the two massive monitor barns. That one would take thirty-six cribbing towers to our four, plus 168,000 pounds of steel beams to elevate the 500,000-pound structure while the floor is plumbed and the foundation rebuilt. But the West Monitor Barn that had been here was entirely dismantled, timber beam by post, slate shingle by clapboard, and rebuilt from the ground up.

The twins, both soon to be in use by the Vermont Youth Conservation Corps, will be reunited at last.

The reconstruction of the West Monitor Barn preceded my presence here, but Art has shared his stories of living in its shadow, which entirely blocked our south-facing house from the morning sun. The sheep used to seek its shade at midday, and Art's cat climbed to the top of the monitor roof and could not find her way down, twice. I like to imagine her up there, her silhouette against the sky like a feline weathervane. After the barn was gone, the house was exposed to the sun and the road, and there was no shade for the sheep in their south pasture until we planted trees there. The bank against which the barn had rested sloped like a pitched roof, and in time, what Art didn't yet know, an invasive thistle would spread over its old footprint, and beyond.

I have seen photos of the barn before it was rebuilt; one graces the cover of the *Field Guide to New England Barns and Farm Buildings*. An old worn-out barn like this with its wrinkled sheathing and sunken backbone has the appearance of something animal rather than built. It wheezes like an old horse. It just wants to lie down, like our old ewes in their final days, and never get up again.

The 1997 coffee table book *The American Barn* also has photos of the two monitor barns as well as a collapsed barn in our town, on our very road. It is not an unfamiliar sight. Art remembers riding past it one day when it was standing, and finding it gone when he rode home. The image says something about how we take care of our farmers, how we take care to remember and acknowledge our history (we don't).

For the magnificent West Monitor Barn, the Richmond Land Trust stepped in just in time.

The banked high-drive barn was not uncommon in New England, but a high-drive, gravity-flow, banked monitor-roofed barn was and is a rare thing. Hay was unloaded from horse-drawn wagons on the high drive, dropped down to the hay mow and into the mangers below, where the cows were milked, and whose manure fell into the basement, where it could be shoveled out. No elevators required. The banked design kept the cows cool in summer and warm in winter. Built at the far end of the horse-powered era, this state-of-the-art technology would soon be obsolete, the wooden multistoried barn with its haylofts to be replaced by concrete-floored single-story structures with manure gutters and silos. Sanitary codes no longer allowed cows on wood floors and required separate milk rooms and later the bulk tank; timber framing was replaced by stick building, and slate roofs and wooden clapboard by asphalt or concrete-and-asbestos shingles. The hayloft, superseded by the grain silo and the plastic-wrapped haybale, would be the stuff of nostalgia now.

It was Moses Whitcomb, who bought some three-hundred-plus acres of abandoned sheep pasture with his father, Uziel, in 1871, who built the two monitor barns, the first in 1901 and the second three years later, as well as our horse barn, the cob barn, and two other horse and carriage barns that are today part of the VYCC campus. The Whitcombs expanded their acreage until they owned "nearly the entire valley" according to the book *The American Barn*, until they lost the farm to the bank in the Great Depression. After a series of failed attempts to make the farm profitable again, Xenophon

Wheeler, who bought the farm in 1948, at last succeeded at reviving the dairy operation, bringing the barns up to code, producing some sixty thousand pounds of milk a day at its peak in the 1960s. You can read the history of agriculture in New England in the story of this farm: the shift from sheep to dairy cows, the advent of mechanization and consolidation, and the great crash of the 1980s when small farms throughout the United States went under in one bankruptcy after another. This was when Wheeler gave up his efforts to keep the dairy going—having lost a not-insignificant part of his meadowland to eminent domain for the construction of an interstate highway—and sold off his land in parcels. Hence our three-acre lot.

Before it dawned on us to give our farm the tongue-in-cheek name Ewetopia Farm (it turns out we were not original in this), after we'd already given generations of ewes *Ewe* names, beginning with Ewedora, Ewenice, and Ewegenie, Art still called his farm the Venture Farm, which was the name of Wheeler's four-hundred-plus acre dairy operation. I never liked the name, which I associated with the French word *ventre*, for stomach, and the even less poetic name for one species of capitalism. It was a source of confusion, too, because the 147 acres owned by our neighbor was also called the Venture Farm, listed in 1994 on the National Registry of Historic Places, though without our two historic barns included (one of them is the only one still housing animals and storing hay), and without the proudest historical landmark of all, the West Monitor Barn. Our Ewetopia is the inverse of a Utopia, which translates from the Greek as *no place*, that is, an ideal that exists nowhere on this green Earth. We begin with a very particular place, with its collapsing barns, its weeds and imperfect house, and from there our imagination goes to work. An empty barn demands animals. There is already a hayloft. We would learn to live with the roar of the road, lest our lives become too idyllic, a reminder, if we needed one, that the world is headed for a cliff.

When Building Heritage was finished, we were left with an empty shell with a dirt floor. In spring Art went to work putting up clapboard, installing floor joists and rough-sawn floorboards, doors and windows, though we would leave the transom open until August for the fledgling barn swallows to leave their nests. This had been their nursery for many years. We might have called it the Swallow Barn.

By the time Art had the studio ready for our use, and had dragged out the electric baby grand he found at a garage sale years ago, the garden was planted. The ewes with their lambs were well into their second course of wildflowers. The swallows were building their nests inside, dozens of them, as they always did. We hung kites overhead, bundles of grasses and flowers for drying, coiled vines for wreath making. On the walls, old farm implements, like the crosscut saw used to cut trees that were as big. And scythes, the rims of old wagon wheels, the horseshoes and old glass bottles that turn up when we dig. Besides a music studio, the barn will be a bicycle mechanic's shop, a potting shed, and a fiber studio; it will be a space for hand-dyed yarn drying on racks, and for felting rugs and spinning; for curing onions and garlic. It will be a yoga studio and a guesthouse.

I would have mixed feelings about evicting the barn swallows, but the barn would be for either their use or for ours. Barn swallows are one species that has expanded its range and numbers with the growth of human populations, who nest under our eaves and bridges, and almost never will do so in anything that isn't human-made. They are a species of "least concern." I knew they would be alright without our barn.

To my surprise, in reading about old barns I learned that the swallow's love for barns was requited, and that there was nothing too progressive about my affection for them or my desire to co-exist. Old New England Barns (but not this one) often had "swallow holes" cut into the gables in decorative patterns. That is, the swallows were *encouraged* to nest inside the barns. As one Connecticut farmer wrote in the *New England Farmer* in 1855:

> In barns built after the old style, "swallow holes" were
> always to be seen. In some of these barns I have counted

twenty nests at one time, all of them being occupied. A
barn swarming with such a multitude of such happy,
innocent inhabitants, resounds with such flutterings,
twitterings and gushing outbursts of song, that it seems
as if everyone who enters within its precincts, even if
he be a confirmed hypochondriac, must forget all his
troubles, and feel his heart drawn upward to praise
Him "to whom praise alone is due," for their cheerful
melodies.... And, besides the pleasure we receive from
their society, they, and especially the swallows, destroy
during their short stay with us an innumerable multitude of insects, which is a fact of no little importance
in these insectivorous times.

That they were encouraged for practical purposes—*a fact of no
little importance*—is not as surprising as this effusive expression of
the farmer's biophilia. Reading this has unsettled my image of the
eighteenth- or nineteenth-century farmer in perpetual war with
wild creatures—this was, after all, the time of the extinction of
passenger pigeons, who were so numerous they darkened the sky
for days on end, and the near extinction, at the hand of the settler
colonist, of innumerable species, among them the big cats, the
wolves, the beaver, and the wild turkey. And yet here is one farmer
in 1855 testifying to the health benefits and pleasures to be gained
from keeping the society of swallows.

Inside our barn there was indeed such a gushing of song. While
we were rebuilding our barn, and then domesticating it by bringing
in flowers and kites and furnishings, the swallows were doing the
same, importing mud from the river for their nests, I am not sure
how. I have seen robins, waxwings, and blue jays carrying building
materials in their beaks—they do like our abundant supply of wool,
and to fray the fibers of baling twine—but I have not seen a swallow
with building material in her beak. We enjoy, together, the feeling of
shelter and security. We are outside of history, or rather burrowed
so deeply into its interior that from here it looks no different to us

than it did in the time of the Whitcombs. We feel *our hearts drawn upward to praise . . . and forget all our troubles . . .*

For a while that summer, until we closed up the last opening, we cohabited. There were dozens of old mud-and-grass nests plastered to the rafters and joists, a few of them still active when we were ready to close up the windows and put in doors. We knocked one down by accident when we were cleaning the inside walls and ceiling of decades' worth of dust with wire brushes. We tried to save it, restoring the eggs to the nest. We found fledglings that had fallen, featherless, who, in their yellowish translucent skin, might have been amphibians, if they were not embryonic dinosaurs. "And so when we examine a nest, we place ourselves at the origin of confidence in the world," Gaston Bachelard writes in *A Poetics of Space.* "Would a bird build its nest if it did not have its instinct for confidence in the world?" As metaphor, *to nest* means to create a place where we feel secure, a domestic space for intimacies and nurturing. (Though as Bachelard observes, among birds, "the nest is not built until later, when the mad love chase across the field is over.") It is also the place where a swallow teaches her fledglings to have the confidence to take to the air, for they are not born with this assurance. And why should they not be afraid? The nest is all they have ever known and it is the antithesis of the open sky. Because a nest is for us the embodiment of confidence, maybe that is why it is so unsettling to see a nest disturbed, fallen from the rafters and undone.

One family of swallows came to be familiar. I counted five baby birds in that one nest, who would peer over the sides of their mud house to look at me and to listen when I sat at the piano. They were curious about me, too, and would listen to me play while the parents flew in and out through the open transom to take care of them without any fear of me. Each time a parent arrived, the little birds lifted their heads up and began a chorus of chatter, an electric buzzing. They opened wide their beaks, showing their ochre-colored mouths, as if that was all there was of them. They transformed into carnivorous flowers, when they opened those beaks, reminding me of the pitcher plants we saw in Newfoundland. Also reminding me of those lonely coasts, with their houses tethered to rock—the great big ball, with its memory of the sea.

We are privileged to know something of the domestic lives of sheep, having observed their nursery, the lambing pen, where a ewe cares for her newborn lambs. But to witness a wild winged creature, a swallow, up close, in her scene of tender domesticity, is a rare gift. For several years, a robin built her nest each spring in a corner of our patio, over an outdoor light fixture. Each year, she built a new nest on top of the old one, until she had a towering column that nearly touched the overhanging eave. I wondered how tall she could build her column before it would topple over. It became a calendar formed of annual rings by which we could measure her time with us. And then one year, she did not come.

To find an old, abandoned bird's nest is also a gift. I have collected a few of them in my time. A nest that fits in the palm of my hand is to me a beautiful thing because it is crafted, woven, artful. The vortices of grasses turned as a bowl is turned, or a lump of clay is centered on a wheel. ("It should be noted," Bachelard writes, "that few dreamers of nests like a swallow's nest, which, they say, is made of saliva and mud.") Because I, too, am an artisan, I feel a kinship with these winged creatures who are so unlike us, but they, too, have a need to nest. The nest is the center holding the world together, it is our source of confidence in the world and in ourselves. The ship's mooring, the great steel ball on a chain, is what tugs at us, pulls us seaward, toward the danger of the breakers and the swells, the least secure places. How we live in the tension between the need to nest and the desire to break free, to risk losing it all.

Once the windows were all installed and the swallows had left, and the *corn crib barn* had become our *piano barn*, we did find one corn cob wedged in a corner under the upper floor, and we have left it there. That and a few old swallow nests.

Later that summer, after we had enclosed the transom, a swallow tried to fly through the glass. She knocked against it, then hovered before the glass looking for a way inside. Art guessed that this was the swallow who lost her nest when we knocked it over, who had come back to start all over again. The next spring our swallows would return to find they were locked out of their old house. Some would make their nests under the eaves, but we never would have such a multitude of swallows, such a gushing of song, in our home after that. For that we would have to ramble out of doors, to venture out under an open sky.

It Pays to Eat Well

As distant as my life on a Vermont farmstead appears to be from my Brooklyn childhood, there are food memories that bind them, and these are among the most indelible of memories.

The first time I visited our new house in Brooklyn with my mother, when I was ten, she took me to eat Lebanese food. We went to a small restaurant on Court Street off Atlantic Avenue, decorated with some light-pink linen tablecloths and some faded photographs of olive groves on the walls. We ate shish kebab and salad, and a smoky goopy dish called *baba ganoush*, with freshly baked Syrian bread. To a palate accustomed to the blandness of American and English fare, of boiled or fried potatoes and naked, unadorned meats that all too closely resembled the body parts they once were, the taste of tahini and pita bread, lemon and olive oil, and honey-soaked pistachio pastries was a revelation.

For many years we ate at this restaurant, which we called It Pays to Eat Well, thinking that was its name, because the phrase was displayed prominently over the door, beside its real but far blander name, Near East Restaurant. Because our move to the new house was delayed, we lived for three months in a single-room-occupancy hotel in Brooklyn Heights, where all six of us shared a single room, the four children sleeping head to foot in one full-sized bed. During this time, we frequently ate at It Pays to Eat Well, which had become our closest substitute for hearth.

We moved into the house when it was still unfinished, and it was a long time before the kitchen was functional. I don't know when we actually began having meals at home, other than pizza from the Queens Pizzeria on Court Street, or spinach and meat pies from the Damascus Bakery on Atlantic Avenue. Often, we ate at It Pays to Eat Well, where the Arab waiters treated us cordially, always with a professional courtesy—no matter how hard my mother tried to engage them in conversation, they remained reserved. She did succeed, at times, in drawing the cook out of his lair and into conversation, an old man with whom she could share her memories of her visit to Lebanon, and from whom she elicited stories about the home country. That was in the early years; at some point, the old man never appeared again, and the waiters would only smile or nod respectfully in response to my mother's questions about their families, or about Lebanon, or the politics of the Middle East.

For years, we never tried anything else on the menu other than what we predictably ordered—the shish kebab, the baba ganoush, the dinner salads, and the pastries. We would read the menu thoroughly, ponder it for a while, and then, one by one, would announce, "I think I'll have the shish kebab." The waiter would write down our orders, as if he needed to. We frequented the restaurant even during the years, after my father had left us, when we often had no money. For all the indignities of our lives—the sibling savagery at home, the unfinished house, the kitchen full of cockroaches—the ceremony of eating out conferred on our lives a measure of refinement. It was altogether unlike the chaos of our home life, which was marked by the unpredictability of our father's rages or our mother's shifts of mood. Sometimes there would be dinner, and sometimes not, but eating in that house was entirely devoid of ceremony and it was rarely an occasion for pleasure.

The seventies passed into the eighties, Nixon was followed by Ford, who was followed by Carter and then Reagan, and at some point, we stopped ordering the shish kebab. We still began with the basket of fresh bread, the baba ganoush with its puddle of olive oil and its heavy dusting of paprika, but since we had discovered the kafta kebab, which we now ordered with the same predictability as we had once ordered the shish kebab, we never touched that old standard again.

To get to It Pays to Eat Well from our house we had only to walk a few blocks. We had to pass the Brooklyn House of Detention—a massive cinderblock tower that occupied an entire block, and then the St. Vincent's Home for Boys. Atlantic Avenue was another world altogether, with its Middle Eastern groceries holding the archives of eleven thousand years of agricultural history on display, beginning with the ancient grains of Mesopotamia—barley and cracked wheat—and continuing through the galleries of time with every variety of olive, legume, and tree nut known to humankind. I liked to enter these stores, to hear the peal of the bell on the door as I went in, just to experience the almost numinous clouds of scent—cardamom and cumin and saffron and honey. They had the feeling of a market in Marrakesh, crammed and disorderly, in which every square inch was filled—sacks of coffee beans (green) and pistachios and semolina flour piled up from the floor, shelves stacked with boxes and tins, and at the highest level, abutting the ceiling, were the hookahs gazing down on us like prehistoric birds. There were giant glass jars filled with rose petals and lavender, boxes of teas and cans of stuffed grape leaves, enormous architectural mounds of halva, pastries, dates, and dried apricots. A scruffy cat was always present, perched on a flour sack, or poised sphinxlike between crates of unidentifiable brown roots and cakes of hand-made soap. To visit Sahadi's, or any of the other Middle Eastern markets, was to pass through a magic curtain to a time before smells were locked up in cellophane, before the sound of the scoop in the grain bin was banished in favor of the safety seal, before the gritty sensuality of seeds and roots and oils were all exchanged in favor of the laboratories of the flavorists, the barcode and the brand name, and the sterile Euclidean geometries of the American supermarket.

In my twenties, I became a vegetarian and discovered the world of vegetarian Mediterranean fare at It Pays to Eat Well—falafel and hummus and tabouli. I often went there with my mother, just the two of us, at a time when we were not getting along. While I was still living at her house, unhappily, we would sometimes go there after a bad argument. As a kind of olive branch, she would invite me to go to dinner, and we would go to It Pays to Eat Well and sit there quietly, like an old married couple that has run out of things to talk about, but at least, over pita bread and baba ganoush, Turkish coffee and baklava, we would establish an insecure peace.

Once, my sister and I arranged to meet our mother at the restaurant after we had gone to the polls during a presidential primary. Even though my sister and I were living elsewhere, we still returned to our old neighborhood to cast our votes. That day, our mother walked out on us in the middle of dinner, after we told her we had voted for Jesse Jackson. Her children had turned out to be such a disappointment, she announced. None of them were interested in music, and even her son, a reporter who covered the White House, who was friendly with Ronald Reagan, had voted for Jesse Jackson!

Even during this episode, the waiters were impassive; none of our odd behavior seemed to raise any eyebrows. These same waiters who had known us since we were small children, who had watched us grow into young women, costumed in the dubious styles of the eighties, with our shoulder pads and big hair—they still acted as if we were strangers, standing aloof and poised with their pens over their pads as we pondered the menu, as we ordered our tabouli and bird's nests and cardamom-scented coffee.

The last time I ate at It Pays to Eat Well, I was with my brother while our mother lay paralyzed from a stroke at the nearby Long Island College Hospital. It had been a long time since either of us had eaten there, and I was knocked over by waves of nostalgia—the same tablecloths, the same food, the same waiters. We were estranged with grief, sitting at a table where we must have had hundreds of meals, most of them with our mother, and we tasted the food that I associated more with her than the foods she cooked at home (she wasn't a cook). "Do you remember?" my brother asked our waiter (the same brother who was rumored to have voted for Jesse Jackson in 1984, who was friendly with Ronald Reagan). "We used to come here all the time?" Our waiter nodded and barely cracked a smile—perhaps he didn't remember us at all and was only being polite. "This was our mother's favorite place." As if our emotions should have been reciprocated, as in a reunion of old friends after a long separation. Surely he would remember her, the one who always tried to engage them in conversation, who treated them as if they were distant cousins she wanted to get to know and to pull into her fold, the one who prattled on about politics, who would have to search through her immense pocketbook to find her checkbook, sometimes needing to spread its contents out on the table, which she did without a trace of shame.

After she died, I didn't visit Brooklyn for many years. When I did return, in 2001, I was astonished to find the neighborhood vastly changed. In the twenty-four years between 1968, when we moved to Brooklyn, and 1992, when my mother died, nothing had changed—not the empty lots, not the corner bodega, not the long shadow of the jail cast over our homes. But in the decade since, all of Smith Street and Court Street had been gentrified, the bodegas were gone, the *abogados* and foot clinics and bail bondsmen, most of the Middle Eastern groceries—all replaced by sushi restaurants and panini shops and boutiques. There was no more wafting of the scent of fresh-baked bread from the ovens of the Syrian bakeries. I would have to walk up and down Court Street, between Atlantic Avenue and Pacific, several times before I would at last accept that It Pays to Eat Well was gone.

Mesclun Days

It was the year the cold war ended, the same year I stepped out of one life and into another. The time when it also "occurred to me while picking beans, the secret of happiness."

The first time I visited Blooming Hill Farm, a young woman with a red bandana over her long chestnut hair was hanging sheets of pasta dough in the summer kitchen. Flour dusted her hands and arms up to her elbows. It was midday on a Saturday and the farm crew was gathered in the summer kitchen, while the pasta artist, whose name was Anne Marie, lay her ribbons of dough on a board and cut them, then pressed them into butterfly shapes. *Farfalle.*

It was a hot day but it was cool in the shade of the trees that surrounded the kitchen. In my memory I hear windchimes mingling with voices as a cool breeze drifts through the trees. The summer kitchen, as I remember it, was the orbital center of the farm, which felt as if it were the center of the world. It stood on the hill overlooking the rows of flowers and early summer greens, and the vineyards in the distance, surrounded by woods, with the rim of the Catskills rising above them. One giant old walnut tree guarded the entrance and shaded the stone steps descending the slope from the kitchen to the gardens.

When I came and went from the city as a weekend volunteer it was as if I was a bird who had alighted on the rim of the world and I was looking in. I may not even have been noticed. Over the next few weeks I would drift in on the weekend and then leave, and then drift in again a week later. Anne Marie and her group had already been living and working together on the farm for several weeks and had developed an intimacy. I watched as the young women played together in the summer kitchen like kittens, stroking one another and purring and climbing into one another's laps. The guys were shirtless and never took their boots off and strummed on out-of-tune guitars. No one understood them when they spoke.

On that initiating weekend I used a hoe for the first time in my life, and I learned what a weed was. I picked baby lettuces and all kinds of greens I did not know the names for but soon learned: arugula, radicchio, frisée, mizuna, escarole, mâche, baby chard, tat soi, purple mustard, claytonia, sorrel, purslane. We gathered what we picked in lugs that we then dunked in a water trough and then lifted and let the water trickle back into the trough. That was before we washed them in great big tanks, picking through the leaves for any bits of grass or yellowing leaves or debris. Much of what I did that summer was pick lettuces and greens and wash them. The owner of the farm, Guy Jones, used to say that he introduced mesclun to New York and that was true enough. None of us had ever tasted greens like this, raised as we were on iceberg and, sometimes, romaine. Anne Marie said washing lettuce wasn't really farming, it was restaurant work, and there was some prescience to her saying that because in time, long after the price of mesclun crashed from sixteen to four dollars a pound, Blooming Hill Farm would host a restaurant to stay in business, but for Guy Jones it was always the golden beets and broccolini, tomatillos and heirloom tomatoes, that were the point.

I think I was a little intimated by Guy, because he was a big man with a big voice and he very much projected his authority as the boss of the farm. The farm was not a commune and it was not a democracy. Guy had a wavy mane of blond hair he often tied back with a bandana, or it fell to his shoulders from under a wide-brimmed hat. I always knew he was near when I heard the squeak of his rubber boots and the rumble of his wheelbarrow. I learned much later that in his former life he had been a labor attorney

who helped street kids and had volunteered on the Cesar Chavez–
Dolores Huerta campaign for farmworkers in California. Chavez
required his volunteers to experience work in the fields, and this
was how Guy discovered he liked farming. I think I would have had
less skepticism about him had I known this, but in time I came to
see him as a fair man, and I remember the farm as a place where I
experienced a freedom I had never before known. He would take
a demolition job if he needed to in order to pay his workers, and I
don't think any of us really appreciated how hard this was to do, week
after week, in times of drought or too much rain and especially after
the price of mesclun fell through the floor—the mesclun that was
worth every penny because that tasteless stuff that would in time
be found in plastic tubs in every supermarket could not compare
to the greens we planted and tended and so lovingly washed. He
paid his live-in crew one hundred dollars in cash at the end of every
week, which to me seemed a meager wage, but then by the end of
summer his crew members left the farm with a pile of cash they
hadn't touched and it added up. For what was the need of money
when you had everything you needed to live?

When we were not working in the fields we were in the summer
kitchen most of the time, cooking foods foraged or cultivated, baking
bread, eating omelets with wild spinach and goat cheese and later
zucchini, and even later leeks, and on Sunday mornings we baked
biscuits we ate with raspberry jam or we baked apple cake. I learned
to cook nettles and lived on salads made of poetry—*claytonia,
arugula, mâche*—and learned to decorate a cake with flowers and
wild amaranth. The path from farm to the great big table in the
summer kitchen could not have been shorter.

"A bowl of beans, rest for tired bones," Kristen Kimball writes at
the end of her memoir, *A Dirty Life*. "These are reasonable roots for
a life. They have comforted our species for all of time. Cook things,
eat them with other people. If you can tire your bones while growing
the beans, so much the better for you." Robin Wall Kimmerer says
the reason a bowl of beans can be so deeply satisfying is because
we feel the land is loving us back. I have struggled with this idea

that "the earth loves us," as she believes. But I don't know what else to call this reciprocity of nurturing and being nurtured in return, this feeling, if it is not a kind of love.

At the time I was living in a second-story railroad flat apartment in Williamsburg, Brooklyn, at the end of a two-year apprenticeship at the Jewelry Arts Institute, a kind of medieval workshop in the middle of Manhattan's Upper West Side where we specialized in ancient arts of jewelry making. I took a job that summer working for a model maker for a commercial jewelry manufacturer who designed and made faux classical models for castings, and I was his only assistant. The studio was hot and without air conditioning and so was the subway ride. The model maker seemed to dislike the work even more than I did—at Jewelry Arts we looked down on casting, and to make a model in crude metals with its clunky sprues never did give you the satisfaction of seeing a finished, polished piece of fine jewelry. The model maker counted the minutes until he could rush to catch his train to New Paltz at the end of the week to spend his weekend swinging from a rope swing. All week long he dreamed about that swing. And I dreamed of Blooming Hill.

I took the job because I needed work, not knowing that learning to make castings would serve me well, or that I would end up designing a chain for one of those pieces that only I knew how to make. I was paid by the inch and got very fast at it, and by summer's end I was earning enough money making that chain that I was able to go live on the farm, returning one day a week to solder and weave a rope swing's length of that chain. Sometimes on market days I left the farm before dawn and rode with Guy in his loaded truck to the Union Square Greenmarket, where he sold his produce, to return with him at the end of a very long day. I crept into the cabin where Anne Marie was already asleep—I was living in the cabin now—and I would sit for a while on the doorstep in the moonlight, letting my mind catch up with my body that had just whiplashed from one world into another.

Later in the summer and into fall the pages of my journal filled up with heaps of globe amaranth flowers, which we loaded into the Vermont Cart to carry to the hay barn to hang to dry with all the flowers and grasses we picked all summer: the fescues and teasel and millet, rye and wheat, the statice, straw flowers, hollyhocks, delphinium, and larkspur. Also Queen Anne's lace, and wild sorrel, and later staghorn sumac and the dried husks of milkweed flowers. We coiled grape vines into wreathes. I climbed up to reach the rafters and came down with my hair full of hay and seed and dried petals, sweaty and smelling of dried grasses and dust.

When I brought my friend from the city to the farm and showed her the hay barn full of flowers and grasses hanging from every rafter and post and joist, summer's harvest still holding on to its ephemeral beauty, all the colors shifting from summer to autumn to winter hues, she said, "It looks like an art project."

9:30 pm. Writing by candlelight. Today we were picking peas when we were interrupted by a group of customers. They spoke in accents, asking "How much? How many? We pick 'em? You need experience at this? Herbs? What is herbs? Parsley? Oh. We put it in soup. How much? A little, just a little." They taste the snap peas right off the vines. A young woman in a yellow dress keeps saying she's never tasted anything like it. "Delicious!"

. . .

Nighttime in the summer kitchen, alone, writing by candlelight. There is a hint of light in the sky behind the dark rim of the Catskills. The windchimes float their notes like falling water. The cicadas surround us with their song. A light shines from the house, the cabins are already dark where everyone has gone to bed. I remember where I was last night: sleeping on my kitchen floor with one fan blowing at my head and another at my feet, the subway rumbling and screeching beneath me. How cool it feels up here, up on the hill.

Before I discovered all the pages and pages in my journal I scribbled while at the farm, I held on to a few strong and vivid memories over the years. Picking lettuces for hours on end through a week of uninterrupted rain. The chill on my fingers from picking frost-covered mint. The hay barn full of bundles of flowers and grasses hanging from above. I remember when I learned of the end of the Soviet Union, perhaps the only time that news of the outside world filtered in through the insulating music of the crickets that enveloped us. What surprises me now is that I wrote often, on my lunch break, at dusk, in the evenings. Any chance I had to snatch a pen and scribble.

Dusk. A misty light hangs over the meadow where the horses graze, drapes from the vines like cobwebs. I watch as the horses disappear with the daylight. The moon had moved behind the cabin, a three-quarter moon. Two nights ago Anne Marie had me step outside to view a bright orange moon, sliced precisely in half like a fruit. Was that the night the cow was loose? When we heard her munching outside our door and found her romping through the cilantro beds. Anne Marie chased her and she actually jumped and I caught a glimpse of the moon beneath her.

. . .

At the end of the workday I was headed toward the summer kitchen when I saw Willy and Kevin arrive. They sat down with a case of beer and some Chinese food and turned the radio on. I wanted to get away from the noise and so I went for a walk all the way down to the stream where I found the cairn Drew had made. I waded in the cool water for a while and when I got back they were still there and so I joined them. The Germans were there too, who I had promised dinner. They were talking about Bukowski, Dylan Thomas, Brautigan. "We've got to defend the peace," Willy said, recalling the time when he was tear-gassed. "We stopped the war." The Germans are drinking and brooding, too. "There are three things that will destroy a man," Manfred says. "Too much beer, too much love, and too much thinking." Willy laughs. We laugh, Manfred too.

Later. Feeling a little sentimental about the departure of the Germans. I made them a breakfast of biscuits and omelets with fresh herbs, peppers and tomatoes. We exchanged addresses and said our goodbyes, and now

I am alone on the farm with Willy. I liked Martin the most, the way he would stand there, holding an egg in the air between his thumb and forefinger and ask, "Anybody want an egg?"

In those days I carried around two books by the poet Richard Hugo: his little book of craft essays, *The Triggering Town*, and his *Selected Poems*. I read them on the bus to and from Monroe; I read them lying on my bed in the cabin, by candlelight. It was his poems about Italy that I returned to again and again, as well as his essay on the origin of those poems, *Ci Vediamo*, about his return to Italy twenty years after he was stationed there as a bombardier in World War II. There are two places he wants to go back to when he visits with his wife in 1963: the site of his former base, and a field, just an empty field. A field of wind. In his poem "Spinazzola: Quella Cantina La" ("I was farther than that farm where the road/slants off to nowhere, and the field I'm sure/ is in this wine or that man's voice") he circles around and around the memory, trying to find that field, to flush the right words from out of the long grass.

I can't explain why I was so drawn to the poem. The way that the language took root in a particular place. Then and now I admire how subjectivity in Hugo's poems is a force strong enough to alter the dimensions of space and time. The field is in the wine. No—the wind is in the voice. As someone who led such an interior life, I held to poems that reflected the power of feeling to warp time into space, and the poet's ability to make a language entirely his own.

I can't have been compelled by Hugo's interrogation of memory, which was not yet an interest of mine at that age, nor could I have related to the experience of returning to a place that has so thoroughly changed after many years. Italy in 1963 was a place "full of fountains, shiny little cars and well-kept streets," he writes. In 1944 "it was a sad place." He wanted it "to be sad once again." Now I understand how it is we want the places where our most powerful memories are rooted to remain the same.

I have not gone back to Blooming Hill Farm. Soon after I left I heard that Guy lost his agreement with the landowner to take care of the vineyard in exchange for use of the land, the house, and the cabins, and had to move his farm. I don't know how you transplant a whole farm, but to me I could not imagine Blooming Hill without the summer kitchen and the stone steps leading up to it, the big old walnut tree, the stone house and the cabin that seemed to float over carpets of cilantro and basil in the moonlight. I came upon an essay almost twenty years ago now in the *London Review of Books* written by someone who apprenticed on the farm three years after I was there, and then returned twelve years later to find Guy's three kids working the till in a farm store and in a café offering brioche and biscotti and bread from Balthazar's. A farm store. A till. Today the farm has a farm-to-table restaurant serving four-course pescatarian dinners with wine pairings and wood-fired pizza with hibiscus pickled ramps or spinach garlic confit or nettle pesto. The store offers swag for sale. T-shirts and ball caps. I am happy that Guy has succeeded in keeping the farm afloat and that it supports the livelihoods of his three sons and a whole contingent of restaurant, farm, business, event, and store managers. But I cannot go back. It was in my time a field of lettuces and flowers, a time when we lived in spartan cabins with only our sleeping bags, a change of clothes, and our journals we wrote in by candlelight. I want it to be so once again.

I know that Guy Jones's intention was not to create a retreat for discontents but to produce food that people would buy. Honest food. Farming was, in his words, "a righteous occupation." The farm was not just a place for dreaming but a business that needed to survive by selling mesclun and dried flowers in Union Square every week, and if I saw Guy and his wife and young son going out into the fields, carrying baskets, to pick their vegetables for dinner, there was also much I did not see. I found out from the essay in the *London Review* that Guy's wife had left. That was after Blooming Hill Farm had to pick up and move elsewhere.

I have no photographs from that time, as I did not carry a camera in those days. I have no photos of Anne Marie, or the summer kitchen, or the rows and rows of cosmos and nasturtium and delphinium, the stone house or the vineyard. Susan Sontag said that the need to take photographs was a need to have reality confirmed, but reality for me had never been more confirmed, and what I hold on to now are the liquid images in my memory and the pages in my book. I have had no reunions with anyone who shared those days with me, not with Anne Marie , who I am unable to find, nor any of the others whose surnames I never even knew. My friend who visited me from the city, who thought the hay barn with all the flowers looked like an art project, does not remember this nor would I remember her visit if I had not written this in my book. I could go back to visit Guy Jones, and a grown-up Travis Jones, who is a toddler in my memory, but that, I fear, would wreak havoc with Blooming Hill as it lives in my memory and imagination, and I admit that the two are easily blurred.

After we finished our day's work, Anne Marie and I sat down in the grass and I recalled the morning, when we crouched down in the wet field to pick lettuce, wrapped in sweaters, and I said to her, I can hardly believe that was today, not just because the day was so long but because of all the changes, how by midmorning we were in the sun—no more sweaters—bent over the nasturtiums, the rows of cosmos floating their orbs of pink, white, and red over our heads . . .

Say it was all a dream. Say it was never as innocent as all that. Say the cow never did jump over the moon in the cilantro beds or the tomatoes did not taste as sweet. Say that the apple cake you baked was not as delicious as you remember, or the omelets with leeks and the biscuits with raspberry jam. Say the frost that turned the mint a moonlight blue did not vanish with morning sun like an apparition. Say you never loved a place so much. Say you were never as free as you were then.

After I went back to live in the city, I returned as a volunteer to make dried flower arrangements, as this was the only work on the farm at this time of year. After making jewelry for so many years it was liberating to work with flowers. I stood in the barn surrounded by dried flowers and grasses we had brought down and stood upright in baskets, and I reached for a handful of statice and then turned and grabbed a few stalks of millet or the dried husks of milkweed. I used my whole body in a kind of dance with the material as if it had as much say in my choices as I did. I loved the feel of working with organic materials that gave themselves freely, that did not need to be wrested from the earth and then heated, cooled, and heated again, to be wrangled into form. I needed no tools, other than a pair of shears. I was a part, in other words, of an art project, one that began many weeks back when someone planted a seed, or in midsummer gathered wild grasses from a meadow and bundled them and hung them in the hayshed. And maybe I would be the one standing in Union Square on a cold December day when this bouquet or wall hanging would be the one I would pass on to a customer, who would carry it away and down the subway steps and then up to her fourth-story walk-up in Brooklyn.

I can see now that the life I had in the city was a full one. My name was on the lease of a rent-stabilized apartment in Brooklyn only two subway stops from the East Village. After a succession of incompatible roommates, who lived awkwardly with me in a railroad flat with little privacy, at last I found one who was as at ease with this and with me as I was with her. I was at home in the dusty medieval jewelry workshop with its torches and pickle pots, its lapidary wheels and rolling mills, its oddball craftspeople, as I was never at home in an office, and my apprenticeship had given me a trade and the freedom that comes with that. Besides, I was a volunteer editor for an art journal that covered the Lower East Side galleries, I reviewed books for *Publishers Weekly*, I took poetry workshops at the Y, which came with swimming privileges, too. I read voraciously, I saw new plays at Playwright's Horizons and the Public Theater. I volunteered in a homeless shelter. I saw the Frido Kahlo exhibit at the Grey Art Gallery, the paintings of Anselm Kiefer that I would never forget, though I can't remember where, Joseph Cornell's boxes at MOMA. I was an obsessive runner who ran thirty miles every week who had the good luck at last to be working near Central Park. I was finding work making jewelry, and could have built a career as a craftsperson had I stayed. In the final months before I quit the city, I put my knowledge of casting to the service of making a collection of inexpensive silver earrings and rings, which is what I envisioned when the dream of making jewelry had first come to me. I had finally escaped the cubicle and the mind-numbing corporate job. And yet I would leave it all, carrying with me only my craft and a few tools, to chase after the drift of mist on a hill, the scent of lavender and smoke, the lightness of being I had come to know when I carried only a sleeping bag, a change of clothes, and two books.

It will be so hard to leave this place, I said to Anne Marie, as the nights grew longer, after a first hard frost marked a hard transition from summer to fall, when we no longer heard the cicadas and the geese flew over signaling that winter was on its way and nothing would ever last.

That winter I bought a one-way ticket to London and volunteered on homestead farms in England and Scotland, and by summer I was shepherding goats in France. Nowhere I went would replicate what I had found on Blooming Hill. Not for many years would I eat omelets with leeks like those I tasted in the summer kitchen, apple cake or biscuits with raspberry jam *en plein air*. I have, ever since, been chasing the scent of paradise I lived at Blooming Hill. I would have found my way without having found that place, that hill, it's true. Still, it set a compass course for my life, giving me my true north, which, in time, would lead me here to this one.

Shearing Day

*M*ay 2009. Yesterday was shearing day. The sheep have all had their haircuts and now they look like a flock of strange goats.

Every year, Jim McCrae and Liz Willis come sometime during the month of May or early June to shear our sheep. I don't have a lot of experience with this, but Art has seen a number of shearers do their work—none, he says, as expertly as these two.

It was a beautiful day—the first sunny day after days of rain. The lilacs and apple blossoms had passed, the roses were in full bloom, and the grass had already gone to seed. The sheep were corralled in the barn when the shearers arrived. Jim looked them over, counted them, and then said, "I think we'll do it right in here." They squeezed through the gate with all their tools, changed into their shearer's booties, plugged in their electric razors, and went right to work.

We are often asked why we don't shear our own sheep. The answer is that we have seen it done clumsily. Sheep shearing done by apprentice hands is torture all around—but especially for the animals. It is uncomfortable for them—the only blessing is that it is fast. Or—it should be.

The first time I ever saw it done was in 2005, when Art's regular shearer showed up on crutches. We knew he was in recovery from cancer treatments—but the crutches were a result of a car accident, not his cancer. He pulled up in a beat-up blue van with his son—a big blond young man in a ball cap, who held his arm to his chest as if it were broken—and a friend whom they'd enlisted to help them out. The son had been knocked over by a ram earlier in the day and hurt his arm, he told us. The old man emerged slowly from the van with his crutches, stood himself upright, then inched his way toward the paddock. He had been completely paralyzed after the accident, he told us. "I've been getting my movement back little by little."

Standing in the paddock, the son lit a cigarette and held it in his left hand while he held his right arm against his chest.

"Go get that one," he said to the friend, who slipped into the corral and grabbed a ewe by her wool.

"Ride her," the son said, and the friend climbed onto the ewe's back and rode her like she were a bucking colt in a rodeo. Then he jumped off her, grabbed one of her legs, and flipped her onto her back. The son wrapped his good arm around her neck and held her against him as he worked the electric razor down her belly with his injured arm. He winced with pain, then stopped.

"You hold her," he said to his friend. The old man stood by, muttering advice with his head downcast as if he were unable to lift it up. He held himself still as a post, leaning on his crutches. The ewe writhed and kicked as the friend held her down, while the son made repeated stabs at her with the razor. Finally the ewe settled down, stretched out on her side, while the son winced and clenched his teeth as he worked. When he had finished the one side they turned the ewe onto her other side. The son worked slowly, using his left hand, the wool peeling away and dropping to the ground like wood shavings. After about a half an hour, they released the poor girl, who went limping off toward the corral to join her family, covered with red nicks and gashes.

It took them three hours to shear six ewes. We paid them eighty dollars to split between them for their efforts, and we never saw them again.

The following spring, it was our neighbor, Jennifer Gilligan, who recommended Jim McRae to us, who has been shearing our sheep ever since.

The two of them worked side by side, each of them shearing their own ewe. Working in concert, they dove into the flock, who were massed together in the back of the barn, grabbed the first two ewes they could get a hold of, and swiftly, using a few expert Judo moves, had them sitting quietly on their rumps. First, they peel away the belly wool, which is discarded, working carefully around the udders and sensitive parts. Then they cut away the fleece in a single piece, carving it as if it were soft butter. The underside of the fleece is astonishingly clean, soft, and beautiful, piling up in folds like whipped cream.

The sheep sit like humans sit in chairs while they have their haircuts, with their backs upright and their legs in front of them. They look a little stunned as they are forced to sit in a position that they would never get themselves into on their own. They seem helpless and vulnerable, with their pink udders exposed. They don't like it—they dislike being handled, they are uncomfortable, and they seem spooked by the buzz of the electric razor. But I have a feeling that they like the sensation of the razor against their skin. They love to scratch themselves against trees or fence posts, and for the duration, they are calm—until the razor is switched off and they get their toenails clipped, whereupon they resume kicking and writhing.

It was Liz who sheared our ram—a 275-pound behemoth. Dressed in his big black coat, Oliver has the bulk of a black bear. Liz is probably half his size. Jim looked over his shoulder with his eyebrow raised as she rassled with him. "You need some help with that guy?" he asked. "I'm good," she said, calmly. She grabbed him by one leg, twisted his head slightly, and had him down on his rump. He writhed and kicked for a moment or two, then he settled down.

Not everyone likes shearing day as I do. An old-timer interviewed in the *Foxfire* books says he "despises" sheep shearing. I can see why—he recounts how the sheep were laid out on a table with their feet tied together. Using traditional shears, it took half an hour to shear a single sheep. Recently I talked with a friend who said he and his wife tried their hands at raising sheep. They read books, took workshops, built a barn—it was the shearing that did it for them. "We hated it," he said.

But I have always liked this rite of spring. It is true, the sheep are not as attractive without their wool: freshly sheared, they look like naked old men, with their knobby knees and their various lumps. In their wool coats, they're each of them individual and identifiable—Ewelysses and Euphrates, Ewe Two and Ewekelele— each of them has a distinctive shape, shade, and personality. But sheared, not even the lambs can recognize their own mothers, and for the rest of the day, they cry out to them like lost orphans. Still, I like to watch the practice of a skill well done, especially a skill as ancient and relatively unchanged as sheep shearing. To witness the magical metamorphosis of nothing but grass into the most useful fiber humans have ever known. I like to watch the sheep step away from their robes as if they were stepping out of the shower, naked to their toes and a touch embarrassed. I love it that we can reap a harvest from our flock without harming them, year after year.

All in all, Jim and Liz sheared ten ewes and Oliver and were finished in less than an hour. They left us with twenty-five to thirty pounds of wool, which will be scoured, carded, drafted, spun, felted, and woven, knitted, or crocheted into sweaters, mittens, hats, scarves, blankets, purses, and novelties. No microplastic. No chemicals. Wool is fire-resistant (firefighters wear wool clothing) and resistant to mildew; we sleep better under wool, which lowers our heart rate and helps us to regulate our body temperature. That much-loved wool sweater can have an afterlife when it is upcycled (as were Bernie's famous mittens) or hooked into rugs. And at the very end of its life, it can be composted to feed the soil from which it came, keeping the cycle going forever. Wool is the sun's energy bottled and saved, no fossil fuel required. Jimmy Carter was famously scorned for telling Americans to turn down their thermostats and put on a sweater. How different the world would look now if we had listened.

Food as Medicine

The Farm at
Vermont Youth Conservation Corps

Around nine in the morning I see a train of plainclothes volunteers walking toward the western fields, carrying totes. When I join them about an hour later they are finishing up the beet harvest. It is a crew from Ben & Jerry's, which regularly supplies volunteers to work on the educational farm next door. They are teasing baby beets from out of the weeds.

Olivia, who is supervising, announces that it is time to move on to the leeks. She instructs them to consolidate their bundles by packing fourteen bunches to a tote. Olivia is a tall twenty-four-year-old who wears her long brown hair pulled back behind a pink headband and has the wholesome, sun-tinted complexion of someone who works outdoors. It is her job to supervise the volunteers, some of whom would not know a beet from a potato.

As the Ben & Jerry's crew finishes packing the beets, I see another group of volunteers has just arrived. There are girls dressed in brightly colored headscarves and flowing skirts over their jeans, and boys dressed in T-shirts and long pants. Students in a summer English language learners program at Winooski High School, they are new immigrants from Somalia, Kenya, Burma, Vietnam, Nepal, and Iraq. Many of them worked at the farm for six weeks in the spring in an afterschool program, caring for seedlings in the greenhouses and planting spring onions and potatoes in the fields. They know their way around the farm. They are moving toward a weedy patch of potatoes when Paul Feenan, the Farm and Food Program director, catches up with them.

He yanks a tall weed out of the ground. "See this weed? he says. "It's called *velvet leaf*. It's a vile and pernicious weed. *Vile* and *pernicious*—there's two vocabulary words for you."

The kids laugh. "Vile! Pernicious!"

"If you see this weed—pull it out!"

Feenan greets his alumni with friendly hugs. "We're so excited to have you guys back," he says. "Now we're ready to harvest the potatoes you planted."

"Aloo!" they cheer. Everyone here seems to know the Nepali word for potato.

Feenan shows them what the potato plants look like and instructs them to pull out the weeds crowding around them. "It will make it much easier to dig them up," he says. The teacher, Karen Green, admits she "wouldn't know a potato plant from a hole in the wall."

"Is this a weed, or a potato?" someone asks pointing to a leafy plant. They go to work yanking up lamb's quarters and grasses as tall as the potato plants.

The youngest girl in the group, who came here from a refugee camp in Thailand, is standing between the others as they weed, taking a very long time to find a song on her cell phone. She is small and skinny for a twelve-year-old, dressed in pink and turquoise and sequins. At last she finds her song. She dances and sings out loud. I ask her what the music is. "I Want You Back," she says to me, smiling. The others join in with her.

When I moved here in 2007, we were surrounded by nothing by industrial corn. That was not long after the Vermont Youth Conservation Corps moved its headquarters into the newly renovated West Monitor Barn that had been relocated from our place down the road and rebuilt. Little by little the corps developed its farm program, reclaiming the land from the decades of tilling and chemical poisoning. Only a few years after they planted their first gardens, the farm was cultivating eight acres of vegetables in a season and distributing fifty-three thousand pounds of fresh produce to food-insecure Vermonters through a health care share program, in partnership with area hospitals.

In the beginning, when those first gardens were planted, I was rather doubtful. I watched as the seedlings struggled in the compacted and exhausted soils. A strange yellowish grass plagued the land, outcompeting every other weed. I watched as a storm, barreling through the wind tunnel formed by the river valley, ripped apart their first greenhouse—from my window I saw its canopy roll across the valley like a tumbleweed across the plains. A year later, the organization hired two experienced organic farmers to teach them how to grow food and to get their farm up and running. They turned the cornfield into a pasture where they grazed Black Angus cattle for a season. A year later, they planted their first rows of carrots, beets, lettuces, and onions.

That afternoon, inside the structure they call the farm barn—a single-story modern building that houses a walk-in cooler, the farm stand, and an array of farm tools—rows and rows of opened boxes are lined up across the floor. So far, each one contains a few yellow and green squashes and some cucumbers. Bruce, Stephanie, Olivia, and Ethan are distributing the harvest to the boxes, while Joe and Tracy weigh and bag green beans over a table. Behind them is a whiteboard on the wall with a chart packed with numbers. Joe writes down "464 lbs" in the "cukes" column, and "448" under "squash." He records a total of 912 pounds for the morning's squash and cucumber harvest alone.

Once all the boxes are filled, the lids must be closed and all 220 boxes must be carried and stacked inside the cooler. When the last few boxes are sealed and put away it is precisely five o' clock—the end of the workday for the crew but not for the apprentices.

All those boxes will go to patients identified by their health providers as in need of access to healthful food. Physicians testify that the impact of the health care share program has been unequivocal. The impacts on the youth who participate at the other end—who plant and harvest and distribute the food—is also undeniably positive, even for those who have learned that they don't particularly like to farm. They are learning what it means to be connected—to their food, to the land, to one another, and to the broader community. Farm crews include new Americans who have come from refugee camps on the other side of the world, who harvest and plant side by side with fifteen-year-olds who have never traveled outside the Green Mountain state, who discuss social issues over lunch, cook giant meals, and break bread together.

"In midsummer," says farm apprentice Cae Keenan, "the farm is a beautiful sight to see. It's also a vision of something greater—the combined work of hundreds of people in a celebration of food."

Lunchtime is the time when the crew has its Writing, Reading, and Discussion (WORD) session, a staple of all the corps programs. The kids, who are all between the ages of fifteen and seventeen, take turns selecting a topic from an anthology of short articles on a variety of social, political, and environmental issues. The core philosophy of the Civilian Conservation Corps created by Franklin Delano Roosevelt in the 1930s—on which the Vermont corps is based—would anticipate the principles of the educational reformers of the 1960s, who sought to abolish the separation between the theoretical and the practical, between learning and doing. The corps is not a vocational training program but an educational one that treats the whole human being, who is learning cross-cultural communication, how to debate and problem solve, how to show up on time, and that onions really do come out of the ground.

On the day I sit in on a WORD session, before they begin, farm crew leader Jeremy Schleining requests that someone explain to a teenage girl, a volunteer, who is the only one not wearing the corps uniform, what they are doing. The volunteer looks at the book, titled WORD, and asks, "Is this like church?"

"No, it is not at all like church," Jeremy says. Then he goes through the rules: everyone must be at eye level, only one person can talk at a time, and the facilitator cannot take a side in the discussion.

Stephanie leads today's discussion of the article she has chosen, "Due Process and Equal Protection for Gays and Lesbians." She reads it aloud. Then she reads the first question for discussion: "Why do people get married in our society?"

The teenagers are slumped in their chairs, while one of the Nepali girls, despite the rules about being "at eye level" is stretched out on the floor.

"Because they love each other," Tracy volunteers in a sarcastic tone. Others echo her response. The whole topic of gay marriage is entirely uncontroversial. There is nothing to debate. Why not? they say. Many of the kids are quiet. It is hot, and the kids are tired after working in the sun all morning. I think of those Millet paintings of peasants napping in haystacks after a long morning of scything.

When they have gone through all the questions, Jeremy instructs them to write in their journals for fifteen minutes. Sumitra is still stretched out on the floor, her arms adorned with bangles. Jeremy looks over at Bruce, who is doodling in his notebook, not writing. He sees that Jeremy is looking at him.

"I'm empty," he says.

In the afternoon, the volunteer crew from Ben & Jerry's has a discussion of their own inside the Hay Mow in the West Monitor Barn, led by one of the farm apprentices, Will Lintilhac, on the topic of civic engagement. The Hay Mow is the one floor of the barn that looks just as it would have a century ago, when the hay was brought in on horse-drawn carriages and then dropped down to the stanchions on the lower level. It is all wood—exposed timbers, rafters, and sheathing forming a latticework of timbers and catwalks that extend to the monitor roof, where the light pours in from the sky and falls to the floors below in long dust-filled columns.

"For my parents' and grandparents' generation," Will begins, "civic engagement was a measure of success. We need to think about what it is that is keeping us from being more involved."

He asks for volunteers to take turns reading selections from Bobby Kennedy's speech on the Four Dangers, which he had typed up onto index cards:

"**First is the danger of futility**; the belief there is nothing one man or one woman can do against the enormous array of the world's ills . . ."

"**The second danger is that of expediency**; of those who say that hopes and beliefs must bend before immediate necessities . . ."

"**A third danger is timidity**. Few men are willing to brave the disapproval of their fellows, the censure of their colleagues, the wrath of their society . . ."

"**For the fortunate amongst us, the fourth danger is comfort**; the temptation to follow the easy and familiar path of personal ambition and financial success . . ."

Will asks the participants to introduce themselves by naming a community to which they feel connected. The scoopers are mostly young—many of them fresh high school graduates who are on their way to college—dressed in B&J's bright tie-dyed shirts.

"The high school community."

"The college community."

"The mentoring community."

"The knitting community."

Only one among them cannot name a community to which he feels he belongs.

A young man with a beard offers, "The bearded community." We laugh.

Will is dressed in a faded green youth corps shirt and a ball cap, his long strawberry blond hair falling to his shoulders. He has blue eyes, a bewhiskered face, and large calloused hands. His brow is often knitted.

"Back in the day, life wasn't so structured. My boss, Paul Feenan, talks about how when he was a kid, he used to get together with his buddies to go play basketball. They would find the key to the gym and just go. That doesn't happen today. A bunch of eleven-year-old kids can't just go get the key to the gym. Someone has to organize it and pile the kids into a minivan. It's very structured. You could say, we are the minivan generation."

Will breaks us up into two groups and asks us to brainstorm "themes and opportunities" for civic engagement. Our group comes up with this list, which is written in black markers on a large conference pad:

Voting and political activism
Refugee integration
Environmental conservation and sustainability
Online social interaction
Education—awareness
The arts
Food—community meals and gardens

To the right of the list is written the word *revolution*, which floats in the white space like a bird circling around in the sky looking for a place to land.

For much of my adult life I have been an educator—through advocacy or through classroom teaching. In my community college research writing classes, students researched and discussed the global environmental crisis, world hunger, resources wars, and the effects of globalization on local economies. These are overwhelming subjects. I could see the students build confidence in their ability to navigate a confusing media universe, but I could also sense how they felt crushed by the weight of the world's problems. I could sense their anxiety and confusion. They could feel—but could not articulate—a dissonance between their sense of entitlement and the limited opportunities before them. They often expressed their feelings of helplessness. What is to be done? That should not be a paralyzing question when there is so much to be done, starting in one's own community. To pick up a shovel, to grow food for the hungry, to rebuild exhausted soils, to repair an eroded riverbank. This is what can be done. If the world is going to be generous to us, then we will have to take care of it.

As a college student myself, I did not feel overwhelmed by the world's problems, as these rarely came up in the classes I enrolled in, at a time when discussions about "language" (few of these discussions were actually about language) were all the rage. In the tradition of an Occidental liberal arts education, I studied the Greeks, the mystics, the poets. I fell hard for the myth of the solitary thinker, who does not see that she is *thinking with* layers of rootedness—in place, in culture, in family, in those who came before. "Such a solitary mind—if it could exist—would be a boring prisoner of abstractions," Gary Snyder wrote. I had allowed myself to be lost in abstraction, so much so that I lost my footing altogether. I think it all began with Plato, who seduced me, I realize now, because he was such a gifted poet and dramatist, though he had it reversed—the shadow world was not this one, but the abstractions he called Forms.

I could have used some of Orwell's roses, with their anchors in the particular, in the world of time. I know that at that age, I would have benefited from working a shovel or holding a handful of seeds. I could have used a sense of purpose and agency. "Work without thought is drudgery," was a core principle of FDR's conservation corps. And thought without work, without roots in the particular and the concrete, is in danger of drifting into the shadow world of abstraction, or worse, as Orwell warned, into rigid ideology.

And I have seen how creative writing classes can serve to ground students in the truth of their own stories, to draw them out of the theoretical realm where they spend so much of their lives as students. Nothing I have seen, in the classroom, is more confidence-building or generates as much joy as writing poems. Even just learning to keep a journal, as the corps crews are taught to do. They are learning to give things their proper names: A rose. A velvet leaf. A silver-spotted butterfly.

One day the following spring I wait with farm apprentice Nicole Mitchell beside the greenhouse for the afternoon crew, the English language learners from the Winooski High School, to arrive. We've already selected the flats of red onion seedlings, dunked them in fish emulsion, and laid them out for the kids who will plant them in the fields. Nicole is a bit fidgety. Looking over toward the back of the West Monitor Barn for the crew's arrival, she says, "I hate waiting."

"Ah!" she exclaims finally (after we have waited for a long three minutes). "They're here!" We watch the group appear over a rise with the magnificent barn behind them. The students form a circle around Nicole, who gives them their instructions for the afternoon. "Today is a perfect day," she says. "The volunteers who were here today did an amazing job planting onions. I know you can work even faster than they did."

Nicole—who studied anthropology and Chinese in college—is beginning her second summer in her food security VISTA position as a farm apprentice. Lucky for Nicole, who hates waiting, she doesn't often have time on her hands with nothing to do, because the farm apprentices are up at daybreak and generally work until nightfall, six or seven days a week. They not only have to dig, weed, and plant but also manage teams of fifteen-year-olds, lead them in reading and discussion sessions over lunch, teach cooking classes, and coordinate sometimes large groups of volunteers who often can't distinguish a potato plant from a parsnip.

We follow Nicole, letting the onion flats hang down by our sides as we walk toward the rows prepared for the seedlings. Jeremy, who was last year's summer farm crew leader, and Nicole demonstrate their method of poking holes in the plastic weed barrier using a spacer. The plants are then gently pulled from the flats, dropped four to a row onto the plastic, and then planted. Crouched down over the ground, we loosen the clumps of clay soil and press the young onions into the earth.

Jeremy and Kadu, who are working beside me as we plant, are deep in conversation—something about religion and Robert Frost. A high school junior, Kadu is one of Vermont's many new immigrants of Nepali descent who were forced to leave Bhutan and then were not welcomed when they tried to rebuild their lives in Nepal. I remember how Kadu, who worked on the farm crew last summer, was shy and insecure about his English. Now he talks excitedly, leaping from one topic to the next as we move down the row.

When we have planted nearly three-quarters of the row, about the length of a long city block, Nicole thanks the group for their hard work and tells them they can take a break before their bus arrives to take them home. "You don't have to stay," she said, "but I'd really appreciate it if some of you volunteered to help me to finish this row." Most of them—and all the boys—vanish before I can even turn around, but three of the girls have volunteered to stay— Fartun, who is from Kenya, and Rosie and Zuti, from Thailand.

It has been a long time since I have tried to maintain that crouched position for so long, and my body is screaming, but I also stay. It is my chance to visit with some of the girls I met one day last summer. Rosie, wearing a bright blue headscarf, jeans, and silver cowboy boots, remembers me from that day. We had lunch together in the Hay Mow, when she explained to me so patiently why she and her friends were not fasting although it was Ramadan.

I recall how Zuti told me that she came here with her large family from a refugee camp in Thailand where they lived for eleven years, after the family was forced to flee Myanmar. She wears a long plaid skirt and a red headscarf that frames a refined, melancholy face and falls down over her shoulders. She loves to farm, she told me on that midsummer day. "I plant all this," she said, gesturing with a sweeping motion over the rows and rows of squashes, tomatoes, peppers, onions, potatoes, and leeks. She is surprised that I remember her. But how could I forget?

The five of us work quickly and finish the row. "I am so happy that I don't have to stay until dark finishing this row all by myself," Nicole says. "Now you better run to catch your bus!"

The three girls run off, and Nicole and I walk back down the row, where we come to the property line that divides our place from the educational farm. I am already home. Over the following week, I will pass by and admire the long rows of tender onion seedlings, forming ribbons of green across a growing tapestry of row covers and spring plantings.

Next year, the farm will produce even more food, to feed yet more hungry families. Aside from all the beautiful produce, I know that by far the most important product of the farm is the impact on the lives of the youth who pass through here. As I leave Nicole and head home, I can't help but feel that, after working for a few hours on the farm, a little of that magic has rubbed off on me, too.

Spring into Summer

When the first warm days of spring come, at the end of sugaring season, when we no longer need to keep a fire going, and the syrup is all jarred and put away, I start to go out into the world on my bicycle, or with Art on our tandem. The world has opened up and it's as if I have grown wings, as I fly through the landscape, along the river, past hay meadows and hills just stirring from their winter slumber. The river shows through the trees that have not yet leafed out; only a faint russet color is cast on the bristly canopy with the white capillary lines of the birches, like a blush coming on. The river is high and fast, forming crescents of white water where it rushes around boulders just barely submerged, the light clear as spring water. The first wildflowers to appear are the coltsfoots, growing along the edge of the road, their stems hooked like shepherd's crooks. Over the next days and weeks, a powdery lime green mist will sink into the hollows as the aspens unfurl their leaves, then a lush velour will blanket the hay meadows and pastures in an emerald green, never as brilliant at any other time of year. Then the buttercups appear, floating yellow orbs over the feathery tops of the grasses. From their patios or gardens, bent over flowerbeds, spades in hand, people wave to us as we pass, especially if we are on the tandem, we are such a surprising sight. Giant weeping willows drop garlands of yellow pearls, and pink beads are forming on the apple trees. Then come the dandelions, crayon yellow, pouring over the green fields thick as honey. Apple blossoms waft cottony in soft green canopies, then paint the ground in pointillist dabs.

The lilacs come and go. Some days, when rain is imminent, the silver maples along the river turn back their leaves, as if to listen. One day, I arrive to see the light on a freshly mowed swath curving its way down the knoll behind our home. I think, Olmstead could not have made a better design. Russet blurs into mint into egg yolk yellow and then the cottony down from the cottonwood trees and dandelions gone to seed. I must have blinked when the leaves came out, but here they are, already on their way to summer, holding autumn's reds and golds within, like banked fires that will in their time set the world ablaze.

Asterix

As farmers, we are not supposed to be sentimental about our animals. That becomes a challenge with the bottle lambs—those lambs unable to nurse, or who are abandoned by their mothers, who we scoop up into our arms and care for like they were our babies. Asterix was a bottle lamb, born to a yearling ewe in April when the sheep were out on pasture and the dandelions and daffodils were in bloom. Sometimes a yearling ewe-mom will take good care of her first born, but that is rare.

Named after the Belgian anti-imperialist comic-strip hero who belonged to a small insurgent group of Gaulois who resisted the Romans invasions—Asterix became our pet. Espresso-and-cream-colored, he followed us around everywhere, bounded up our front and back steps, and if the door was left open, he would prance right into the house, where he would dance around, springing into the air and tapping his little hooves on the floor, his little tail trembling.

Lambs grow up to be aloof, not so cuddly, and easily spooked, even the bottle lambs, but Asterix grew up to be a gentle wether, who liked to stand beside us and have his chin scratched like an old golden retriever.

We always intended to give him away. It was only after he bumped me a couple of times that we decided it was time. Then one day he rammed me in earnest. He came at me from behind and knocked me halfway across the paddock—where I landed in a pile of shit. Then he backed up to come at me again. And again.

That was it.

In the beginning of our project as sheep farmers, in our innocence, we hoped to intervene as little as possible in the lives of our animals. That meant not separating the ram from the flock. There are good reasons, though, for farmers to time the rutting season—so the lambs aren't born in winter but in spring. There are also good reasons for castrating the ram lambs (an altogether painless process involving the simplest technology—a rubber band)—otherwise you will not only get lambs in winter but you will get inbred sheep. We discovered, though, with Asterix, that neutering the ram lambs doesn't make them less aggressive. The older they get, the meaner they still get. Sometimes our interventions in their lives do actually minimize their suffering. The sheep are cooler in summer without their wool coats, for instance. Interventions in their "natural" lives can also minimize our own suffering.

Rams, in short, are dangerous animals. That ram, born during the blizzard, was sweet until he was suddenly mean—when he attacked Art one day and broke his ribs. That was over Labor Day weekend, the same weekend that our cat died. In the morning I found Art trying to dig a grave with his freshly broken ribs, using only one side of his body, after a sleepless night spent sitting up in pain. "I'll do that," I said.

Bottle lambs can turn into the most dangerous of all. They have absolutely no fear of us. After Asterix, we would no longer keep a ram at all. And farm animals, even the bottle lambs, are not our pets.

There's a story by Naomi Mattis in a favorite anthology of mine, *Intimate Nature: The Bond between Women and Animals*, about the author's encounter, while walking alone at the edge of a national forest in New Mexico, with a ram. She had heard talk about this ram, "who didn't seem to belong to anyone and lived wild in the hills." One day she came face-to-face with him on a deserted road. The ram stood firmly in her way and she needed to get past him. When he backed up, she thought, *Oh good, he's leaving.* Anyone who is familiar with rams will know that this is *not* what this means. "Instead, he lowered his head and charged toward me with full force." He knocked her to the ground. She got up, he backed up, and he knocked her down again. And again. He rammed her so many times it's a wonder she didn't break any bones. Finally, she came around beside him and grabbed onto his fleece and held on to him from his side as he moved, the two of them circling round and round this way, *for hours*, until she was found and rescued. "The ram insisted," she writes, "that my hands remain on his body." There is nothing mystical in this—a ram's only weapon is his head used as a battering ram and this he cannot do unless he comes at his object head on. His head cannot be twisted or he will break his neck. If you are beside him, rather than in front, he cannot harm you. The friend who saved her from this dance wanted to kill the ram but Mattis had no desire to see him killed. "The ram was my ally," she writes. He was only doing what he had to do as was she, and while they were bound together they were also bound in their willfulness. But he was not her ally. This kind of aggression is a consequence of domestication and is not a characteristic of the species' wild progenitors. No wild animal would ever do such a thing. (An animal that wanted to eat you would come at you with stealth.) A ram loose like that is a grave danger. She is lucky that she lived to tell the tale.

That afternoon we offered Asterix as a giveaway on Craigslist and right away we received several responses. Two days later, two large men showed up in a pickup truck to collect him. I had spoken with a woman on the phone, who told me that she and her husband raised goats and that her husband "just liked to watch the animals out on pasture." I warned her that the ram was aggressive. Her husband could handle that, she said. When the man showed up, I was a bit relieved, as he was a giant of a man who was easily a match for poor Asterix.

The man lifted Asterix onto the truck single-handedly and sealed him in a plywood box. The other man just watched. Neither of them looked like farmers. They did not ask any questions about the animal they had just adopted; they did not even ask his name. In truth, they acted as if they were taking him straight to the slaughter.

"He will need to be sheared soon," Art said.

The man looked up at Art. It seemed to me that he had no intention of shearing him, and that the only "soon" for poor Asterix would be the butcher's knife. I could hear Asterix through the plywood—his hoofs tapping on the wood—a new sensation for him—in his agitation.

I had decided that if someone wanted to take Asterix for slaughter then I would accept that. The animal posed a danger. Why else would anyone want him? Ewes provide mowing services and wool without the menace. But that night I had nightmarish thoughts about Asterix's journey in the dark box in the back of the truck, having never been apart from his family and having never known any life but this one—here on these few acres. For Asterix the world was these pastures and the old barn, the shade of the elm tree and the willow, the hum of the road and the chickens.

Every winter, as long as we were raising lambs, a few were slaughtered right here on the farm, so that they never had to go through the trauma of being hauled in the back of a truck and then rough-handled by strangers at the slaughterhouse. Art was there to say goodbye to them. I stayed inside the house and did not hear a thing, and when I went outside, I would find the flock calmly picking through their hay or chewing their cud, as if nothing out of the ordinary had occurred.

But this, I feared, was not to be the end for poor Asterix.

Rain, Rain

I had lost count of the number of days it had rained. When was the last time we had seen the sun shine? We are lucky to live on top of a terrace. I can look down over the Winooski River valley and watch the river spill over the floodplain, the green mountains behind it, and know that we are above the floodwaters, and yet we still live beneath the clouds, with their enormous loads of rain.

During this long stretch of rainy days and nights, I was once woken in the middle of the night by a thunderstorm. I heard the wind toss a load of rain onto the window and looked out to see a young box elder sapling leaning against the house, shaking its spindly limbs in a demented way. There was something about it. The thunder began to roll, coming in louder and closer; I heard the sky crackle and split, then a deep explosion followed by a torrent of rain. I could hear it gushing from the roof gutters, pelting the windows and the roof.

The storm continued for a long, long time. I worried about the animals, though they were surely safe inside the barn. The sheep are afraid of thunder. Our cats, too, who had no doubt run downstairs into the basement. It was on that night that I felt something had happened, some cosmic switch had been pulled. I was living inside a magic realist novel that began with the phrase: *The rain came for a hundred days . . .*

If there were no date to this journal entry, I would think I had been writing about Tropical Storm Irene, which inundated Vermont on August 28, 2011:

Thursday night we had record rainfall. Thunder and lightning began at dusk. I saw a large animal in the pasture adjacent to us come running out of the woods and frantically race around, as if in a panic and completely lost. I first thought it was a horse but when I got a closer look I saw it was a moose. It galloped toward the road, then turned back and rushed back into the woods.

The storm grew more severe in the night. The sky flashed and rumbled, cracked and split. Thunderstorms are often like this. But the next day we passed the Andrews' field across the road from us and saw that it was flooded. The cornfields by the interstate were under water. It was even worse than the spring flooding a few weeks ago; this time the water levels were much higher. I heard on the radio that night that the Winooski was at its highest level since 1927. Montpelier was especially hard hit. The old timers say they had never seen rain like this—and this was after the last few years of so much heavy rain.

Alan K. Betts, a climate scientist living in Vermont, has warned us to expect monsoon-like rains in Vermont, with prolonged dry periods between them. This is the new normal. I predict it will not be long before this record will be broken again.

It was tempting to use religious language, to imagine some angry Old Testament God behind it all. Or to say, as do the Hopi, that Mother Earth has had enough. But no religious language would be needed because science had defined it all. This was time's arrow, which the modern world had pulled back with all the power given to us from fossil fuels, and shot into the future at galactic speed. Entropy, from all our energy use, was on fast-forward. We sped up the flight of the arrow and now it was headed toward some terrible abyss. We were unraveling the order of the earth like a kitten a ball of yarn. All the magnificent creatures who would vanish—the river dolphins, the otherworldly cetaceans, the elephants, our cousins the gorillas and chimpanzees, the big cats and man-eating bears. And the myriad small creatures who will become names entered in some giant ledger, the catalogue of extinctions—the monarch butterfly, the poison dart frog, the tree frog, the tomato frog, the frog who is invisible, the frog who gives birth through her mouth. The comical puffins, the stoical penguins, the whisker-faced seals.

I say "we" because I have to include myself, as a member of the culture that has done all this, as a member of the generation who has done more damage to the earth than any other in the history of the species, and who cannot say, as our parents' generation could, that *we did not know*.

After I stopped hoping the rain would pass, I went down to survey the garden. I remembered what it used to be like after a good rain, when the sun would come out, and with it the birds and their chatter, and the garden would look so refreshed. I remembered, as if from a place in the future when gardening was in a long ago past, what the tops of the carrots felt like when they brushed against the back of my hands as I thinned them, and the soft tearing sounds of the roots as I pulled them, little orange apostrophes, carrots the size of tadpoles. I remembered the pleasures of hanging laundry from the line. I remembered how the sky would look, blue with mottled clouds casting aqua and purple shadows over the dark green mountains. I remembered it all as if it were a long ago past. I surveyed the poppies, the petals scattered on the ground like torn pieces of tissue paper. I surveyed the lettuces shredded by snails.

We had lost some trees: they looked as if they had been ripped out by cyclone winds. One big box elder had been ripped apart limb by limb and lay splayed over the ground behind the big barn. We still had our solitary elm, one of the few survivors of a plague that took out most of the old elms. It lorded it over all the damage, displaying its broad canopy like a peacock's tail. And we had the two willows we planted, who love to have their feet wet, and so they were no doubt loving it now.

Prolonged wet weather is not good for sheep, who evolved in arid climates. I worried they would develop fly strike, a horrible affliction of flesh-eating maggots that breed in their wet wool. The lambs are susceptible to parasites that are worse in wet years. They can also catch pneumonia. But there was nothing more we could do than make sure they had access to their barn.

I watched the farm crews in their green outfits sloshing along the muddy paths in the rain with their heads down, and felt bad for them. For many it was their first experience with farming. Many of the spring seedlings would have to be replanted; others would be delayed by weeks. Seed corn rotted in the ground.

That was the summer we had to pull up all our beautiful tomato plants because they were afflicted with late blight, which was an epidemic that summer. Late blight is an airborne fungus that thrives in wet weather and is carried aloft by storms. I noticed the telltale dark spots on the leaves, and then overnight, the leaves curled up like burning paper. I had to pull them all up and bag them in plastic to take to the landfill, where so many other gardeners were doing the same. I had never had to do anything like this before. I'd lost squash plants to powdery mildew and had seen my cabbage seedlings shrivel up and die for some mysterious reason. But this was different, to lose an entire tomato crop that I'd nurtured from seed and tended for four months. I had to pull them up when they were loaded with green fruits and just beginning to ripen.

Only the blueberries seemed to love the unending rain; they blossomed and grew into great big clusters of purplish berries nearly the size of grapes.

On the same page as my journal entry on the Memorial Day flooding of 2011, I noted that the lower joint of my right big toe was swollen and tender. I was walking around on the side of my foot. This was the first sign of the spirochete invasion that would spread up my arm, shoulder, and finally my right knee, an infection that began with a tick I never saw. *I have no idea what the problem is*, I wrote.

This was all in the antediluvian days before Tropical Storm Irene. O yes, we remember it now. The rain, the rain, all the rain.

The Monitor Elm

We have some great old trees on our property, but the tall two-hundred-year-old elm tree is the grandest of them all.

It rises straight up from the roadside and then opens up like a fountain in the sky.

I have many fond memories of gathering with friends around a picnic table in the shade of its great canopy. Our sheep like to take their siesta in its ample shade, and in summer its crown is filled with songbirds. We have even seen the pendant nest of the Baltimore oriole suspended from its uppermost branches.

It is odd that friends who have been to our place many times are surprised to learn that there is an elm tree here. That may be because its telltale tall straight trunk is disguised by the maple that grows with it like a conjoined twin.

A few years ago a botanist from the University of Vermont stopped by to study the tree; he told us it might be the healthiest American elm tree in the state. More recently, someone with the Nature Conservancy knocked on our door—he was passing by when he happened to notice our elm. The Nature Conservancy is searching the region for the largest and healthiest surviving elm trees for its floodplain forest restoration project, with the goal of planting seven thousand disease-resistant saplings over three years in Connecticut River watershed.

Only last year another large old elm tree in Richmond succumbed to the disease that has taken seventy-seven million elms since the 1970s. The tree was removed by a sawmill that specializes in processing big old trees. The Tilden Street elm was made into a conference table that now dignifies our town offices and can be admired and used by the people of Richmond for generations to come.

Before I learned about the Tilden elm, I had been rather complacent about our own, believing that if it has survived this long then it was not in danger. Then I learned about the death of another elm, this one the largest elm tree in Vermont—called the Vermont Elm—last November in Charlotte. That same sawmill took down the tree, milled and kiln-dried the wood, and turned it into furniture. The bottom twenty-foot-long section alone weighed twenty-five thousand pounds.

Elm trees once dominated the floodplain forests in New England and were planted along city streets to form living arches, in city parks and town centers. They are fast-growing, robust trees that can tolerate urban environments and all kinds of storms. And they are beautiful.

We did love the elm tree to death, however. Planting rows of elms along city streets, in effect, created monocultures that made them susceptible to epidemic disease. When Dutch elm disease arrived, it swept through these plantations, spreading from treetop to treetop, and then infected elms in their natural habitats as well. Trees that remained isolated from affected trees, and those with genetic resistance, survived. Our elm is one of those.

American elms are still abundant in floodplain forests, but they do not survive to maturity. No other tree has come to take the ecological place of the largest, longest-living tree in the floodplain forest. Elms, with their deep strong roots, kept soils from washing away and maintained the water quality of rivers and streams while providing habitat for osprey, eagles, barred owls, songbirds, bats, and flying squirrels.

Trees have a trenchant psychological power over us that may be difficult to explain but makes sense, given the importance of trees to our survival, and given our origins as tree-dwelling primates. In his beautiful essay "The Brown Wasps," Loren Eiseley writes that he passed his life in the shade of a nonexistent tree—a tree that took root and flourished in his memory as it failed to do in the patch of soil where he planted it, with his father, as a young boy. The house and the street where he lived had both rotted away, but not the memory of the tree, which, he learned many years later, had perished in its first season, just after his family moved away. "It was part of my orientation in the universe," he wrote, "and I could not survive without it."

I know what it is like to lose a big old tree. Before I moved to Richmond, I lived beside the New Haven River in an old house surrounded by large trees that were badly damaged in a late summer thunderstorm. One tall black locust tree lay sprawling across the road, others lost their tops. A neighbor broke a leg when a tree fell on him during that storm, which whipped and twisted like a cyclone. For a long time I mourned the loss of those trees, and I felt that sense of disorientation—the kind that comes with grief, when we stumble about on our sea legs, stunned and unanchored in a world we only dimly recognize.

It was a bit like that for us when the Twin Towers fell—we lost our physical orientation in the city and, in some ways, our orientation in the world as well.

Our farm is exposed to the most violent winds that are funneled through the river valley, picking up strength as they barrel across open farmland. In the aftermath we find tree limbs and branches wrenched from their bodies, or whole trees felled and uprooted. There is some risk, or sentimentality, in personifying trees, but it's hard not to, as a tree's life cycle parallels our own, though they will long outlive us. Gachelon Bachelard writes, "The suffering tree is the epitome of pain." But storm-ravaged trees recover, limbs grow back, a hole in a forest canopy lets in light for the growth of young saplings, and the rotted wood from downed trees is host to whole galaxies of life. This may be where the sympathy between human and arboreal life cycles begins and ends.

A favorite poem, "B.C." by William Stafford, imagines the millennial history to which a single sequoia has borne witness, ("before Jesus, before Rome . . ."). I like to think about the history our elm has observed from its transcendent position in the sky—the disappearance, and recovery, of the forests; the meadows dotted with fewer and fewer sheep but more cows, and now a pox of subdivisions and their cul de sacs. This old farm. The raising of a four-story-tall monitor barn, its slow decay, and its resurgence from its ruins.

Our elm tree does not have a name, but perhaps it deserves one. I would suggest the Monitor Elm.

Our box elders have been split and wrenched and mutilated so many times they are by now shapeless malformed trees, with all their amputations and perversions. The birches do what they are supposed to do—they bend over backward and never get up again. But the elm—the elm lords it over all of this mortality and remains unscathed, with its canopy high in the stratosphere.

As the climate becomes more and more deranged, there will be more storms and more lost limbs. The elm—if it remains untouched by Dutch elm disease—with its immense trunk and tenacious roots, may be poised to weather the storms to come. We will need these giants in our floodplains to hold back the waters of the deluge, and the forests to hold the world together through its pain.

Under the Penobagos

A Land Acknowledgment

I feel like it was only a moment ago when it began: a few pockets of fire in green canopies, the colors catching on as the land is set ablaze with fall colors—carnelian reds, sunflower yellows and golds. We, also, are consumed by a burst of late season energy: we plant garlic and spring bulbs, gather up the last of the harvest, put the gardens to bed, and hurry to stash tools and trellises and stakes before the first killing frost, before the snow comes.

I am thinking about how it was at his time of year, under the Penobagos, the Moon of the Falling Leaves, when the Winooski Abenaki were here. They passed through in their canoes, of course, down the Winozkizibo at other times of year, but it was in autumn that they paused, set up their camps, gathered butternuts and acorn, and hunted bear, deer, porcupine, muskrat, beaver, squirrel, even chipmunk. The furs they made into bedding and the hides into coats and tent canopies. They ate and dried the meat and made tools and snowshoes out of bone and sinew. They collected the cottony fibers of milkweed and spun them into cloth, made eel skin cords and cedar bundles for their games, sewed leather into moccasins and skirts.

It's not been easy for me to find these stories—about the Original People who belonged here, so savage was the erasure by the colonists. In my own time archeologists found, at the confluence of the

Winozkizibo and a tributary we call the Huntington, the charred shells of acorns, butternuts, beech nuts, and hickory; bone fragments of twenty-two species of animal; a nut mortar and pestle; projectile points, broken knives and scraping tools; tools made of quartz and basalt, tools made from river cobbles—all buried under seven feet of gravel and sand. This is the telling record of the arrival of the whites: sand and gravel dumped by the flash floods caused by clearing forests for farmland, erasing the traces, the footsteps, the signature of the Original presence. With the forests gone, so did the People move on, too, meaning they either left or changed their patterns on the land, because there were no more butternuts or oaks, no more bear or beaver or muskrat left to hunt.

The Abenaki call this time, after the arrival of the whites, the Years of Darkness.

The swath.

Here I live, on a small New England farm, with its nineteenth-century barns, old apple trees, and pastures. Generations ago my ancestor Abraham Belknap sailed from England with the Great Puritan Migration, settled in Massachusetts, his sons and grandsons and their sons and grandsons carrying the torch of manifest destiny from Massachusetts to Connecticut to Vermont to Ohio and finally to California, where there was no more west to *go west* to. In the restlessness of settler colonists, they kept moving, and with my own move to Vermont, the ancestral story has come full circle. Robin Wall Kimmerer says that gardening is a way of loving the land, and being loved in return, but I also know that cultivating the land has been a way of violence. What I do with this land I do out of love, and yet I know that this farmland, with its old barns and its cleared land, tie me to my settler colonist ancestors who cleared these forests and left behind that swath—

Seven feet of gravel and sand.

I am trying to learn these names because the renaming of places was a form of erasure, and by continuing to use the colonial names we are continuing to dump sand and gravel on the Original presence. With each colonial name a little flag is planted that says "this is ours."

That I with my husband "own" these three acres is an accident of history, personal and familial, beginning with Abraham Belknap's arrival on the shores of a very old world. I know that the legal fiction of ownership is based on a historic theft, but what it means is that I have been given a great responsibility. I can—I must—harness farming as a force for good, be a part of a movement away from supremacy and dispossession, toward regenerating soils and building community across the species barrier. "The moment," Margaret Atwood writes in her poem, when you say "I own this, / is the same moment the trees unloose / their soft arms from around you . . ."

I often look across this valley toward the river, with Tawabotiiwajo, Saddle Mountain—the mountain we call Camel's Hump—behind it and imagine how it once was, before the road, before the house lots and clearings. The fiddleheads and silver maples still grow along the river banks, as they used to, but the floodplain, once hosting a forest of elm, cottonwood, alder, and butternut, has for generations been planted with industrial corn, the soils turned over and washed away, its gifts of arrowheads and bones long given up and looted. Behind us the hills were shorn of their forests and then overgrazed by too many sheep; by now the wild turkey, deer, and beaver have returned, the woods have grown back; the trees hold on to the soils now, but without the understory—the carpet of trillium, lady's slipper, trout lily, and jack o' pulpit, all the medicines; without the big old trees, without the wolves and big cats. The chestnuts, butternuts, and big elms have all died off; the beech trees are diseased and do not grow old; the pines are twisted and covered in blisters. And now it is the brown ash tree that is dying. The Original People knew how to use fire to control pests and disease, but the settlers did not know this, and maybe now that we are ready to listen it is too late.

Reciprocity, respect, reverence, responsibility, relationship. Judy Dow, our Abenaki teacher, says these were the values she was raised by. They were not the values I was raised by, who was taught to be *irreverent* and was never asked what responsibilities might come with my freedoms, but I am learning. Reciprocity, responsibility, and respect form the basis of my gardening practice, and it is the land that has taught me this.

Behind our home the neighboring farm is growing corn with squash and beans. The squash shades the ground and suppresses weeds to help the corn, the corn provides a trellis for the squash, and the beans sequester nitrogen to feed them all. Some young foresters are trying to manage the forests, forests of dying trees, like the First Peoples did, but without the use of fire. It took us many years to remember these teachings, to come back to what might have been.

I harvest sunchokes, grapes, goosefoot, and dandelion, which grow at the blurred edge between wild and domestic. As do the "wild" apples still bearing fruit long after the settlers who planted these trees are gone, and their stone walls, collapsed and morphed into sinuous organic shapes as if they were made by the movement of ice and root, not by the hands of men.

When Judy speaks about the land it is alive and moving before her and I don't mean trees in the wind and water rushing: I mean the mountains themselves and whole rivers rearranging themselves, sea levels rising and falling and ice dams breaking and the water rushing in. All of this change is recorded in the Abenaki stories— they remember the Great White Bear who ground the mountains down to dust and they remember when he withdrew to his place in the sky; they remember the mastodon and mammoth, and they remember how they mourned their disappearance; they remember their sea faring days, their trade networks linking the Arctic to the Mississippi; they remember the speeches of the great orators at the Great Council Fire. And then the Years of Darkness came, about which they do not speak. That swath.

My ancestors put down markers and drew boundaries as if they could keep the world from moving. They named the new places the same as the old. They said "I own this" and the trees unloosed their soft arms from around them.

When the Moon of the Falling Leaves is passed to the Freezing Moon, the Winooski Abenaki would gather up their pelts, all the hides they had prepared, the dried meat, the bear rugs and bundles of bones and claws to use in their ornaments and games, and with their loaded toboggans would head back to their village at the mouth of the Winozkizibo, to rest for the winter, at least until the Moose Hunting Moon, when they might be back.

Mobility, they knew, was the key to conservation. It is how they survived for generations—by following the rivers, living in concert with the changing seasons. The changing season is never more apparent than now, when the landscape flares as if in alarm, saying: *pay attention.* I saw a bear the other day, who came out of the woods to fatten on apples. And today from my window I watch the wild turkeys, drifting over the corn stubble, just passing through under the Penobagos Moon, returned to the land after a long absence, or not so long in another measure of time.

Disturbance

(The Middle Years)

Dreaming Snow

It had not been a good winter for snow. In the weeks leading up to our night in Bryant's cabin, one of three backcountry cabins leased out by the Green Mountain Club, we had grown tired of picking our way over the icy ground. It was Saint Patrick's Day, a day we randomly picked for our one night in the cabin that we'd been gifted by friend who won it in a raffle. In recent years some of the best snowstorms came to us in March, but it wasn't looking good for us. We were resigned to bringing snowshoes for our overnight stay on the mountain rather than skis. But the as the day approached, the forecast predicted snow. Lots of it.

We stopped at the bakery on the way up for pastries, and at the grocery store for cheese and yogurt and bread. It was raining. We drove up the access road in the sloppy weather as the temperature gauge dropped below thirty-two degrees. From the Nordic Center we picked up a sled, which we packed with most of our gear. It was snowing gently. There were a few inches of fresh snow on the ground already and conditions were excellent for skiing. We climbed up Bryant's trail, a long uphill slog to the cabin, Art pulling the sled while I carried a day pack stuffed with our food and water. My skis gripped well.

The wooded cabin is spartan and clean though it is well worn. A picnic table, a big woodstove, and a spacious sleeping loft. We unloaded and changed out of our wet mittens and sweaty underclothes. We boiled water on our gas stove and made tea before heading out.

We started out on Birch Loop, which was not tracked. I went off-trail to ski over and through the pillows of fresh down, gliding through the parklike woodlands, and then circled back. The spruce and balsam trees hung their boughs until they almost touched the ground, arched like dancers, the needles frosted. Now and then a gust stirred up a whirlwind and wrapped us up in a blinding snow-swirl. We went off-trail toward Cotton Brook, a new area for me, and it was beautiful in the way a poem is especially beautiful on a first reading. We skied through the trees, making fresh tracks, then turned back where it suddenly got steep. It was all untracked terrain and would be easy to get lost, but Art knew his way around, as he'd been skiing and snowshoeing this mountain for decades, leading tours and teaching skiing. I followed him through the trees, ducking under the weighted boughs. The snow was really coming down, my wool scarf all white and feathered. The risk of getting lost in the snow as night was falling did cross my mind, but Art was confident. And then we were back to our own tracks, heading in the reverse direction on Birch Loop again, and there was our cabin. We kicked off our skis and we were home.

The stove was still warm when we arrived, with live coals we revived by stoking the stove with new logs. The cabin came supplied with firewood. By this time the snow was piling up and covered the bottom half of the windows. We peeled off our wet clothes and warmed up by the fire and watched the snow falling.

Now and then a skier appeared. Skiers will always pause in front of the cabin, having come to the end of the long slog up the hill. I watched them look over the trail map nailed to a tree; the snow was so high that they had to bend down to read it. They pulled out a thermos and drank and snacked on power bars. They turned to face the door of the cabin and then turned away. We looked at them, but they did not look back. To our surprise, skiers kept coming very late in the day and even in the night, wearing headlights. I felt like I was looking through a trail camara to witness the activity that we normally do not see, the animals that come out in the night and pass us by as we sleep. They do not know we are watching them.

We melted snow in a pot on the stovetop; we cooked couscous and lentils. We drank wine. We bedded down like kings in our palatial loft, just the two of us in a loft that sleeps six.

The storm came during the night. The wind howled, rattling the windows and rumbling beneath us. It shook our world. The snow slipped off the roof and came down in a thunder that startled us awake. We slept as if with one eye open as we listened to the wind come whistling through the trees and through our dreams and it was hard to know where we were, above or below the surface of our dreaming.

In the morning the snow had walled us in—there were only a few inches of daylight at the tops of the windows, which looked like parfaits in glass flutes filled with cream. I opened the front door and faced a wall of snow up to my chest. I reached out with a pot and scooped up some snow to melt. We would have our breakfast and then we would dig ourselves out.

The first skier appeared early. He paused in front of the cabin and stood in silence, alighted like a bird on a branch. Later a whole group appeared, who drank tea from their thermoses and snacked on power bars and peeled the skins from their skis. Another group. These were people we knew. "Hi Art." "Hi Art." We were at Art's work place, after all.

We packed our things and cleaned up, zipped up the sled and stashed it in a corner next to the cabin, and headed out with a bagged lunch. We climbed up Heavenly Highway, a trail that takes you to the top of the world. The early morning skiers had made tracks for us. The trail climbs over a steep stretch, twists through a narrow passage, and then follows along a ridgeline through some great big old trees. It was a sunny clear day and there was no wind. We descended down into a wide-open snow bowl, then snuck through the trees again. I made telemark turns in the deep untracked snow on the wider trails, but slid sideways or sidestepped down the trickier spots, though Art can ski anything with grace and ease. When we came to a descent that scared me, I made Art find a detour, and then we ended up in the alpine area. To pass from the

Nordic backcountry into the downhill skier's world was like passing through a curtain into another reality. But it was a wide open, gently descending trail with few skiers. You could not ask for a better day: fresh snow, sunshine, no wind, no crowds. The skiers glided along as if on air. Eventually we were at the bottom, deposited in a world of chair lifts and skiers dressed like astronauts, condos and parking lots. We took off our skis and walked to the Nordic Center and ate our lunch there. "You're back." "Not really." We weren't supposed to be. We were tired when we slogged back up Bryant to the cabin to pick up the sled. Art strapped it behind him and I followed him back down the mountain as he made his graceful turns even with his load, which carved out a nice smooth path for me.

The Nordic Center was almost buried in snow. The wide trail to the center was elevated high above the tennis courts, and a snow bank leaned against the side of the building reaching above the roof.

As it turned out, this night was the biggest snowstorm of the year—the night we randomly picked back in May—as lucky as being given the winning raffle ticket. When we got back down to the valley everything looked so beautiful smothered in down, but you could see that it was already melting and would not last. This was winter's last great big sigh. It brought us down gently with it.

Looking at Animals

Lambs in spring. Of course, that is when they are supposed to arrive. Not in the dead of winter. We vowed we would not do that again, and this time, we meant it. To postpone the lambing season—which would naturally begin five months after the first cold nights—until the milder weather arrives requires making sure all the ram lambs are neutered, and either separating the ram from the flock or not keeping a ram at all. Last year we had no lambs, and we had no ram, and in the fall we borrowed a ram (who was not easy to find) for two months, returning him after he had performed his "service." He was a gentle giant with a roman nose and a bell hanging from his neck, which meant we always knew when he was near. His name was Obama. I was not sure how I felt about that name, but soon enough the name affectionately belonged to him. Obama did have a fringe of gray frost to his dark wool, and he had that certain presidential coolness about him. After the tragic events of November 8, 2016, I felt a pain something akin to grief whenever we spoke his name.

When he first arrived, the girls (as we call them) were afraid of him and took off running. Obama ran after them, his bell ringing, but since he had a sore ankle he had difficulty keeping up. This went on for a few days. In time, it was the ewes who were seducing him, rubbing up against him and batting their eyelashes, and he became well integrated into the family. He turned out to be the nicest ram we ever had here; he never bullied the ewes and never so much as suggested any aggression toward us. When we had to get him into the truck to take him home, I stood back and watched—having been spooked by the aggression I have seen from other rams—as two girls from next door helped Art to lift up his front legs, then the rear, to get him in the U-Haul. They were dressed in clogs and colorful scarves. Obama only shrugged as he took one last look at us before Art closed the door and he was gone.

The lambs came five months later, in two-week intervals, starting in mid-March. The first lambs were born—both ewes, one black and one white—to Eweriah, a granddaughter of Ewelysses. They were both strong and healthy. The following day we had the biggest snowfall of the season: twenty-nine inches in Burlington—a record. The next day more pairs of twins were born. And then more. All the ewes, in the end, had twins, except for Eweripedes, who produced one big black ram lamb with a white X mark across his face. Lambs in snow still sounds like winter, but the temperatures were in the twenties at the lowest, T-shirt weather for the sheep, who had no trouble with it at all. No frozen ears rigid as tortilla chips, no cold blue tongues. We never had to bring any lambs inside to warm them by the fire. And no bottle lambs.

We dug out a trail to the barn through the knee-deep snow. I looked back at the barn from the house through a lacey curtain of snowfall to see the newborns, white and black, nudging at their mother's underbellies, their heads disappearing and little rumps sticking out, the tail shaking to say to the world that that all is well beneath its milky firmament. The white blanket of snow covered the bare paddock, with the sheep tucked in as if in a freshly made bed. They will not try to make their way through the deep snow, and so they are confined to the barn, but this is a good time for them to stay home, absorbed as they are in their tender domesticity.

And when the sheep are out on pasture, we get to watch the lambs racing across the brilliant green velvet of new grass. They have long tracts of grass for their races, and we watch them tearing across the pastures in a zebra-striped blur. They will play their king-of-the-mountain game on the backs of ewes, who remain placid, chewing their cud, as the lambs jump on their backs and run across the ridgeline of their spines, then jump off and jump again. Art made them a jungle gym out of some pallets, an artificial mountain for their games, which we can watch from our kitchen window. They kick their feet together in midair when they jump, twisting their bodies like a high diver in her descent. They will run circles around their mountain, then take off in a race across the length of the paddock and back, more lambs joining the throng until they are a full herd galloping across the moors.

In his classic essay "Why Look at Animals?" John Berger reflects on the disappearance of animals from our lives who have been with us as our partners in survival throughout our history. In Vermont, there are almost no cows or goats or sheep grazing on open pastures anymore, though that iconic image persists in our minds as if it were still true. There are two large dairies in our town, though no cows anywhere in sight. On one farm, the cows sleep on waterbeds and are milked by robots. One the other, it is possible to catch a glimpse of their big-boned bodies through the barn door and to see the calf hutches lined up behind it, though not the poor animals that must live their short lives in them. The animals are still here but we do not see them, and they cannot return our gaze.

The writer and ecologist Carl Safina has been looking at animals his whole life. I am not talking about sheep but dolphins and whales, elephants and wolves, crows and razorbills. His beautiful book *Beyond Words: What Animals Think and Feel* contains a wealth of observations of animal life that lend a whole new depth to the question, Why look at animals?

When your dog rolls over for you to rub her belly, or wags her tail and beams at you—is it fair to say that she is happy, or would you be *assuming* too much? Is it possible that humans are so different from dogs or cats or lambs or elephants that we are the only ones who experience happiness, or that other creatures cannot express it in similar ways? Science has collected abundant evidence to demonstrate that animals are individuals—as humans are—that they use tools; that they are aware of the minds of others; that they can use deception and cleverness to outsmart others; that whales, dolphins, wolves, and elephants grieve and mourn their dead, and help and care for others in distress; and that they play, laugh, joke, and cry (elephants even produce tears). We are not anthropomorphizing when we apply these words to animals; these emotions and abilities do not belong exclusively to us. "Not assuming they have thoughts and feelings was a good start for a new science. Insisting they did *not* was bad science."

Jane Goodall was famously scorned for her work with chimpanzees when she enrolled as a doctoral student. She shouldn't have given the chimps names, her esteemed professors told her, and she shouldn't have talked about their feelings or personalities. But this insistence on human uniqueness and difference contradicts what we know about biology and evolution—not to mention what is evident to just about anyone who has ever had a dog. Biology tells us that each newer thing in nature is a slight tweak on something older. Everything humans do and possess comes from somewhere—frogs and chickens have femurs, the precursors to our jointed leg. That is to say, that most of what we possess as a species is shared. Species differ, Safina writes, but not by much.

Anyone who thinks that animals have no consciousness, no language, no ability to plan ahead, no self-awareness or consciousness of death, or that they do not have "theory of mind," meaning that they cannot know that "others have thoughts different than their own," would be astonished by the observations recorded by researchers who have logged countless hours patiently watching them. All animals, it turns out, are individuals—even the squishy octopi are different, one from the other, just like we are—where one octopus will screw open a jar to retrieve its contents, and then rescrew it, another will show no interest whatsoever in the jar. A razorbill knows every other razorbill by name—out of thousands in a colony—and can find his mate among them, without fail. Groupers are clever, and solicit the cooperation of eels at capturing their prey. Jays plan consciously, storing food and eating the perishable items first. Killer whales (orcas) live in social units with their distinct dialects and avoid mixing with other units for *cultural* reasons. There is no parallel for this, that we know of, outside of humans. Above all, the species that is most like us is not a primate but the wolf, who lives in family packs, in socially complex groups, who is as loyal as she is brutal. Wolves will banish one another, defend their loved ones to the death, and lead suffering, tragic lives. Males care for their spouses and offspring for life, and bring food home to their families—only humans do this, and no other species is at war with itself, wolf against wolf, the way we are. "Wolves and humans can understand each other," Safina writes. "That's one reason we invited wolves, instead of chimpanzees, into our lives." Wolves, dogs, us. "We were made for one another."

Near beaches where killer whales hurl themselves onto the sand to drag away thousand-pound sea lions, whereupon they beat them and tear their bodies to shreds—the same whales will, docile as puppies, form a ring around a park ranger who has slipped into the surf in his kayak. Whales will seek out our kind, hanging around boats, putting on a show for whale watchers, seeking us out as playmates and companions; dolphins have escorted lost researchers home, guarded humans from sharks, and mourned our dead. Attendant dolphins watched over the Cuban boy Elian Gonzalez, adrift at sea, nudging him back onto his tube to keep him afloat. Is it possible that whales and dolphins detect a kinship with us that they do not feel they have with flounder or seals—and that it has something to do with our minds, our consciousness, or, dare I use the word, our souls?

Elephants, too, have been known to defend and protect us—showing an empathy and kindness that we surely do not deserve. One elephant carried an injured woman to a safe place, covered her in branches for warmth, and sat with her through the night to defend her from hyenas until help arrived. We have heard the stories of elephants carrying people to higher ground during the Indian Ocean Boxing Day tsunami. Elephants will stop dead in their tracks to avoid hurting a human in their path—treatment they would not give to a hyena or a warthog. What is it that elephants see in us that they give us this special treatment? Especially we who are the only species to massacre their entire families, destroy their forests, and cause them to live their lives in terror and grief.

Not all animals mourn their dead, or experience the profound and inconsolable grief that whales and elephants do. I have seen our sheep step over their dead lambs as if they were nothing to them. And when we take their lambs away, when they are a year old, the ewes that gave birth to them and cared for them so tenderly in their first months do not even notice they are gone. They graze and chew their cud and lounge about as if nothing whatsoever were amiss. I can say this with as much confidence as I can say that they are happy when they show happiness, or that they are bored and unsatisfied when they are pestering Art to move them to new pasture.

I suppose that by raising animals we are engaging in our own *thanatology*, the study of how different species respond to death, troubled not by questions of our own afterlives or preexistence but by our human relationship with animals and our need to understand, and come to terms with, the animal deaths that are all around us. It seems that predators whose lives are consumed by the practice and study of killing, who kill in order to live, almost certainly have a concept of death. Maybe that is why our sheep, who are not predators, show no signs of being aware of death, except that they know to flee as a herd when they feel threatened. But they show no signs of grief at the death of their family members, nor angst as they face their own deaths. They are, it seems to me, at peace with death so long as the herd will live on. By this measure, livestock are different from elephants (who are not predators either) and whales, who are the very embodiment of soul. If our sheep wailed and screamed and shook when their lambs were taken, it would surely not happen on our watch.

To say that they do not miss their lambs is not to say that they are insentient, or that they do not suffer, nor is it to deny that each of them is a *who*, not an *it*. (Every time I type a "who" to refer to an animal, grammar-check advises me to change it to a "that.") It is our responsibility as their guardians to provide them a good life, and that means to allow them to live a sheep's life—to borrow from Joel Salatin, that is, a life in which they are allowed to express their "characteristic form of life." That is their ewetopia.

We are all too familiar with images of nature tooth and claw. But what of nature heart and soul—the myriad displays of intelligence, joyfulness, empathy, caring, kindness, loyalty, even soulfulness that is there for us to see in the animal world if we would only bother to look? Having lived with animals since time immemorial, how is it that we have missed all of this wonder and astonishment? Our understanding of animals is coming around full circle to what the aboriginals always knew, who certainly never doubted that animals had "theory of mind." To accept that all animals have inner lives that are a mystery to us—sheep and cattle, too—is to admit wonder into our daily lives. The world, it turns out, is a much more interesting and mysterious place than the one mapped out by Descartes and his descendants—alive with consciousness and empathy, language and music, and webs of communication across the species barrier that we have barely glimpsed. We need not be so lonely as a species—and surely we would not be as cruel—if only we would pay attention to what is hidden in plain sight.

We are not the only ones looking at our animals, who can be seen from the road by passersby who often stop to admire the lambs. Berger says that the presence of animals gives us a sense of permanence. "Once the animals flowed like their milk," he writes. The ewe that has died "had already lambed her permanence." So it was that the animals who had been with us were still with us in our long evolving partnership in survival. That is why we are grateful to see lambs out grazing, or resting together with the ewes beneath the shade trees, chewing their cud, in their blessed contentment. Or the cows on the hillside pasture behind us. This year they are Jerseys, those caramel-colored cows with the long curly eyelashes and the dreamy eyes of young girls in love. The animals have not disappeared altogether, their presence says to us, the soils have not all been carried away by the winds and the rains in their absence. There are still animals out on pastures, to anchor us on this earth, its axis fastened in its reliable turning, four-leggeds speaking in their strange tongues of this day and forever.

Going Gently

We buried our favorite ewe yesterday. Ten years old, Ewelysses died the way the old ewes always do: she just lay down one day and never got up again. It was in the middle of lambing season, and during her last days, the new lambs were coming into the world, playing and crying and sometimes jumping onto her back. Such is the nature of farming: so much life, so much death.

Ewelysses had these crescent moons under her eyes and she had a sense of humor. Over the years she had given us so many sweet lambs, and she doted on them, even though, a bottle lamb herself, she had no memory of receiving such care herself as a newborn. She could be overeager as a mom, and would jump in and start cleaning up the lambs of other ewes, before she had delivered her own. Ewelysses would have to be separated from those lambs before they started nursing her, for once a lamb's tail has the smell of a mother's milk, there's no turning back.

She was also my first bottle lamb, born on the eve of the Valentine's Day blizzard of 2007. I had to hold her in my arms to feed her during that first night, because she would not stand up to drink, but by the morning she was upright and bleating away, and by the second night she was already bounding out of her box.

On the day of the blizzard, another pair of twins arrived. We had to dig a trench through the snow to get to the barn, which quickly filled up with snow again. We asked our tenant to come out to help us with the stubborn ewe, and waited almost an hour for him to make it from the house to the barn through the chest-deep snow.

Later, when the snow hardened, the sheep could walk right over the top of the paddock fence, trailing their lambs behind them.

I remember it all so well because this was the winter I was traveling back and forth to Washington, DC, to help care for my brother in hospice. These were the two lambs who appear in my poem "Coming Home":

> When I got home from the airport
> (it would almost always be evening),
> after visiting my brother in the city,
> after going in and out of his room
> where he slept, after watching over him,
>
> . . .
>
> The first thing I'd do
> was go out to the barn
> to feed the two lambs, where I'd let
> my body sink down to the ground,
> my back against the wall,
> as the lambs—one black, one white—
> climbed all over me, until they found
> their bottles, which they'd suck
> with a great ferocity, until
> they were satiated, and calm,
> the one resting across my lap,
> sleeping, murmuring.
>
> There I would sit for a while
> in the dark, listening to the slow
> heavy breath of the ewes,
> the ground soaked, through the years,
> with the blood of afterbirth,
> and where, when the old ewes die,
> they just lie down in the straw
> and never get up again,
>
> wanting to remain
> with the animals,
> as the old poet said.

It was my first winter here, and somehow I managed to plant dozens of spring bulbs but I can't imagine when. What I do remember so distinctly is one particular morning, in spring, when we were putting down the floor in the piano barn. I was sitting on our porch admiring the bulbs—Queen of Night, White Emperors, grape hyacinth, all the daffodil—some with egg yolk gold centers and white petals, some salmon-colored, some trumpet shaped, some all yellow, a pale lemony yellow. The lungwort a friend had given me from her garden surprised me with its rainbow of pastels. I remember that day so well, as if I had on that morning broken through the prism of my bereavement, the day that winter was finally behind me. I considered that, a year before, the flower bed was still a woodpile. Except for some daylilies, there were no flowers at all around the house. A year before, the ship's mooring was still a buried secret, the barn still listing and full of junk.

The physicists and philosophers tell us that time isn't real, that if you take the long, deep view, our little world of time, packaged into days, into hours, into minutes, when we zoom far out into the eons, where time folds into space, the world is made stranger than we can know. I don't know what else I have to stand on, if not time, because time is all we have in this life, that *current rushing toward death*, as the poet wrote, I don't remember which one. But what of a woman's body, bleeding every month, until it no longer does? What of her deadline on bearing children? What of the body that must punch a clock on a thirteen-hour workday, or rouse itself at 3 a.m. to start another? What of a brother's cancer diagnosis, given a year, at most, to live? Those days and weeks and months go by, the clock clicking loudly like the bootheels of dreaded footsteps approaching. Before the twelve months have passed the brother will have died, ahead of schedule. He did not even have a year.

And what of the changes in a New England town? The barns that were here last year have burned to the ground or were dismantled, one by one. New subdivisions have popped up on what was open land and the meadows have been turned into lawns. Disease has taken the last of the old elms. And now the ash trees. It is only a matter of time.

Even the dead are affected by time. A death grows up and matures like a person, like a tree. A one-year-old death is like an infant, who needs to be held close, who wakes us in the night. As a death grows older it claims some independence from us. Our relationship to it has changed. On that day, when I sat outside and considered that those flowers had not been there a year before, that such a short time ago my brother was still alive—my mother had died seventeen years before, on May 4th, 1992. Her death was by then a mature old tree, and I had lived in its shade all these years. How different is a mature tree from a sapling.

Our days are the shape of our lives. They are line and stanza, musical phrase and bar. They are the glass cups filled and emptied with light, filled and emptied. Even the most chaotic life is contained in its vessels. The patterns of waking, pouring coffee and brewing tea, washing, dressing and undressing, they are our scaffolding over the abyss. I used to think that the work of dailiness—all those domestic chores—was not as meaningful as my more "serious" work (such an elitist notion), until I encountered the words of Gary Snyder, for whom the repetition we practice in our daily lives—cooking, weeding, splitting wood—have value equal to religious rituals, with the same good results. "Such a round of chores is not a set of difficulties we hope to escape from so that we may do our 'practice' which will put us on a path—it *is* our path," he writes. "It can be its own fulfillment, too . . ."

We can accept that daily life, and historical, geologic, and unfathomable time, can all coexist, being equally true. We live our lives in the tune of one time signature, while in the distance, we hear the faint, faraway hum or echo of another.

And now I come to mark the tenth anniversary of my brother's death, which is also my mother's birthday. An animal's death occurs inside a circle of living and dying. I do not rage against it.

My mother never did get to meet the man whom I would marry. Nor did she see me find my way to Vermont, or become a poet, a writer, and a teacher. I imagine my mother, a Juilliard-trained violinist and avant-garde music producer, a single mother who worked a day job as a typist, in conversation with Art about music, about politics, about so many things, sitting at our dining room table in the house in Brooklyn, my mother doing most of the talking. Playing music had been so important in our lives, yet I did not stick with the violin, which I quit before I finished high school, or the piano, which I started playing in college. Or with the recorder, which I picked up in my forties. Art is the one between us—who had to beg his parents for lessons—who has kept up with his guitar playing, his daily practice, inviolate.

Our lives converged, here in the foothills of the Green Mountains, so far from our respective family trees, having come from such very different worlds. Art was raised in a two-parent household, his mother a homemaker, his father a banker, in a New Jersey suburb in the Watchung Mountains, "the world's smallest mountain range," he likes to say. I grew up in an old Brooklyn brownstone with an open-door policy on guests and squatters, a house with a harpsicord, an electric sitar, and Buddha statues in every room (we were not Buddhist), a house where conversations stayed up all night, smoking and listening to Joni Mitchell's *Blue*. Our mothers did have one thing in common—both refused to spend their days chauffeuring their children around, one of the many reasons my mother moved us to the city, where we could get around on our own two feet, subway tokens in hand. Art learned at an early age to rely on his bicycle, and his thumb, and given that the family lived at the top of a mountain (the world's smallest), he would develop the self-reliance—and the legs—that would one day pedal him across the Himalayas at seventeen thousand feet, and all over the Green Mountain state for many years, in all weather, hauling his trailer behind him.

It was his reading of *Walden*, Art told me the first time we met, that planted the desire in him to one day live in the country, closer to the bone. His love of mountains drew him here. That and his travels in Nepal, which gave him an altogether new perspective on his own life, inoculating him against the contagion of consumerism, however imperfectly. He could see how the hustle to keep up with ever-shifting standards of affluence would exhaust his life in the impossibility of filling an insatiable hunger. We discussed other books we loved in common the first time we talked, and on my first visit to this house, I would find his bookshelves held more books on religion and anthropology than I had read as a religion major in college. He was questioning the Catholicism he was raised by, he told me. For me the study of religion was anthropology, curious as I have always been about the many ways of human culture making. None of those books I read in those classes helped me with the burning questions inside of me. They said nothing about our human exceptionalism, which had so troubled me before I understood that it was all a great mistake, a symptom of our species loneliness. It was Primo Levi's testament to human dignity, the insights of Hannah Arendt, who had much to teach me about human cruelty and the resilience of the human spirit. Neither did the theologians have much to say about the mysteries of time, but the poets, Charles Wright among them, and the novelists, are, admittedly, more than a little obsessed with time:

> The heart of the world lies open, leached and ticking with sunlight
> For just a minute or so.

We can look out our windows to see what the weather will be—wild, surprising, ever changing—the *heart of the world lying open*—watch for how the sunset will ignite the sky before it dims, how pale the wafer moon will be or how thick and viscous its honey light—and it is enough. *The last scud of day holds back for me*, Whitman wrote. *It coaxes me to the vapor and the dusk . . . I bequeath myself to the dirt to grow from the grass I love . . .*

I have tried to create a ceremony for the burials of our ewes. This is when I feel the absence of ceremony in our lives, being without formal religion, and though I have tried to create or to borrow them on anniversaries to honor the memory of the dead—burning yahrzeit candles, offering tobacco—they never did feel true.

"Did you say goodbye to her?" I ask Art when he comes back from burying a ewe in the woods.

"Yes," he says, quietly. I know that what we mean by *saying goodbye* is only allowing a moment with ourselves to feel our sadness at letting go.

We gave Ewelysses a forest burial. Art carried her through a drizzling rain in a wheelbarrow across the rain-softened field to the edge of the woods, where he lay her down on the forest floor. The coyotes know the spot—it is our unwritten covenant with them. In a few days, there will be not a trace of her left. She will become a part of the forest, in the song of the coyote on the hill, in the black streak of a raven's wing in its flight. Other than the snow and the hail and the rain, and her birth—a creature coming into existence from the watery, dark warmth of the womb, like a nimbus star—this will be the most wildness that she will ever know. *The last scud of day . . . coaxes her to the vapor and the dusk.* She will leave a ring on the ground, like a shadow, a wisp of wool. And time will take that, too.

Staying Cool

It has been very hot all week. I go out to the garden only at night to water the young plants. This keeps them alive, but still, they do not get a really good drink the way they do from a good rain. The garlic tops are turning yellow, the pea trellises look unhappy. At night we open the windows to cool the house and in the morning we close them; the house stays cool most of the day this way, but not when it is in the nineties. Not when it is this hot, day after day.

By early evening when it is still very hot I go to the river for a swim. The ride there on my bicycle is tough with no shade along the road, but I endure knowing that at the end I will be able to plunge into the cold river. There is a stretch of the lower Huntington where it is possible to swim laps, and to swim in place in the current where there is a small rapid. I swim through the shallows as far as I can before I am scraping bottom, then turn around and swim back around the giant boulders where the water is deepest and coolest, and into the white water, where I swim in place with my nose upstream in the current like a salmon headed home.

Yesterday the forecast said heavy rain in the afternoon, but the sky was clear when I headed out. It started raining as soon as I arrived. I swam the length and looked up to see the needles of rain etching silver circles on the surface, forming starbursts all around me as by now I was alone in the river. The couple who arrived when I did had climbed out of the water and left. And the small group of bathers who were sitting on the giant boulder in their bathing suits, holding their drinks, scattered and ran to take cover under a large tree. I continued to swim, as I heard no thunder. I saw the group still standing under the tree. When the weather is hot like this it is the only time of day when my head feels clear, and all the tiredness and oppression of the heat is washed away. By the time I get back on my bike to ride home it is still very hot but my body has cooled off and I am in my wet bathing suit under my clothes and in ten or fifteen minutes I am home.

By then it is almost time to open the windows. We pick salad and make dinner—garlic scape pesto over pasta, or omelets with sorrel, maybe new potato salad or zucchini sauteed in garlic, maybe stuffed squash blossoms—and eat outside on our porch, watching a cardinal who has landed in the birch tree, listening to the high-pitched screech of the kestrel who flies back and forth between the barn roof and the tallest limbs of the elm. Or the blue jay who has set up house in our big white pine.

The Huntington spills down from the mountains through a spectacular gorge, where the canyon walls have been scooped by the passage of water over the years, and then disappears as it drops down over a lesser gorge. Before it reaches the valley floor, where it joins the Winooski, there are a few nice swimming holes, as well as lonely stretches of shallow water where on a hot day there will be people sitting on lawn chairs with their feet in the cold water, alone or in groups. We used to go to a spot farther upriver that required walking (riding bikes, in our case) down to the water a way through the woods, where there was a large, deep clear pool and a little waterfall where you could wedge your body between the rocks to be pummeled by jets, and then let yourself be carried downstream. Tall cedars reached over the river from the far bank and shaded the stretch of gravelly beach. There were downed whole trees forming pools of their own that had been yanked up by their roots in some storm with no one there to witness but there they were. We stopped going there regularly after the river changed its signature and the pool was not so long anymore, and then we discovered the pool at the bottom of the river, refreshed by a jet of glacial spring water, a popular spot for good reason and so much easier to get to.

Just above the gorge there is a parking lot next to the most popular swimming hole on the river, but we have never gone there for swimming. There are often swimmers diving into the gorge. I have watched this: someone standing on a ledge where the canyon walls swell outward and the jumper has to fall through a narrow gap between the twisting walls of the gorge and then climb out without getting sucked downriver into the maw of the treacherous rocks. It is sheer terror to watch this. I often see people down below the falls, sunbathing on the rocks, and I am not sure how they managed to get there. Many people have drowned in this gorge: on a placard at the site are the names of twenty-two who drowned there, but since the marker was put up in 1996 many more have drowned. Among those who drowned were rescuers who gave their lives to save bathers who were sucked into the deep gorge. But despite all the warnings that "the current is deceptively strong and fast" people continue to swim there, and to jump.

The last person drowned while she was walking along the edge of the swollen river when she slipped and was carried away. Last summer two women were lazily drifting downstream in the gentle current above the falls and were swept over the gorge but were rescued and survived.

Last year the road to the gorge was closed because a storm on Halloween night carved out a big chunk of the road. To ride the road you had to squeeze around two sets of fences and big mounds of gravel and dirt. The upper gorge was closed to swimming and parking, but people still went swimming there.

The other day there was a family there when I arrived at my regular spot near where the Huntington joins the Winooski. They were camped on a little spit of gravel downriver from the boulders where I sit. A group of children were riding the ripples of white water where it is very shallow, the smaller ones wearing floatation wings, or life jackets, the bigger kids riding on tubes. Their heads bobbed in the ruffles as the current carried them along. The big kids in tubes would get stuck on the rocks and would have to rock themselves to get free. I saw limbs tossed here and there and little heads and bodies tumbling as the current carried them. The current was especially strong that day. Meanwhile Mom was holding an infant on the shore while Dad stood by and at some point got a small fire going. Finally the dad joined children as they climbed along the rocky riverbank again, but this time they kept walking upriver to the next beach. I watched them as they floated downriver, and then got on my bike and went home.

Grass Farming

I used to be frustrated by how much time Art spends moving sheep fencing. It would make more sense (though it would be expensive), I reasoned, to put up a fixed perimeter fence around the property and let the sheep graze freely. "But they wouldn't do a good enough job," was his response when I proposed this. That the sheep have a job to do, in return for the services we provide them (a barn, the ever-replenishing water buckets) was the first lesson I had to learn. It was the reason Art brought them here in the first place: to perform grass-mowing services. After his first summer living here, when he hired someone to mow the grass, he remembered an exchange he once had with his father, who made his kids mow their lawn.

"This will be in your future, too, son," his father said.

Art was determined that it would not.

The next year he picked up six adult ewes and stopped having his grass mowed. And he learned to use a scythe.

The first summer I spent time on the farm I watched the man who would become my husband mow a fenceline with a scythe. His body curved into his swing in a steady rhythm as he moved, the felled grasses cascading behind him like long silks. I had never seen anyone use a scythe before, and knew of it only from my reading of Tolstoy, whose passages on his character Levin mowing with the peasants I first encountered as a teenager living in Brooklyn: *Suddenly, as he worked, he had a pleasant sensation of cold on his hot, sweating shoulders, without understanding what it was or where it came from. He looked up at the sky while whetting his scythe. A low, heavy cloud had come up, and fat raindrops were falling . . .*

The sheep will let Art know when they've grown tired with their pasture. They become restless and vocal. And they will know they are about to be moved when they see Art out there swinging his scythe. For their first days in a new area, they are delighted, so absorbed in feasting on all the new grass and flowers, taste testing every new leaf and blade in the salad bowl, that they show no interest in us if we are near. For the first day or two, or longer in midsummer, the grass is *not* greener on the other side.

Before picking up the movable fencing, the sheep need to be moved into the paddock—an enclosed area that gives them access to the barn—which is the only time they are fed grain. They are so crazy about it that they will follow Art anywhere, for the promise of a handful of grain, which is all they are going to get, mixed with mineral salts in a blend Art calls his "Dorito mix." It is a trick to lead them from place to place, but the old ewes will still run after that grain dish, to the sound of Art calling to them "Come on, girls!"

Once they are inside the paddock fence, Art goes around picking up the netting, post by post, laying it down on the grass in an even pattern. Early on I thought it looked simple enough, so I volunteered to help him with this, but the netting got so hopelessly entangled that I never tried this again. As someone who used to make jewelry, I learned to draw the finest wire and then coil it into spools without getting it tangled—watching Art pick up the netting, in his steady rhythm, then draping it over his shoulder, reminded me of this. It is not a task we can ask our sitters to do for us when we are out of town, for it is a rare person who has experience with moving electric netting. But it is a skill that can be learned.

Art trails the netting behind him, laying it down section by section along the newly cut fenceline, which he has mowed with the scythe, and then goes back to hammer in each post (this part I can do). The cycle is complete once the flock has followed him into the new pasture, lured by a grain dish and his call, "Girls!"

The idea of moving grazers from paddock to paddock is to mimic the migratory patterns of wild herbivores, who moved across the prairies in dense packs, from one patch of grassland to the next. There is a science to what Art does when he moves fencing around: grass can be both overgrazed and undergrazed, and will lose its productivity either way. A blade of grass will recover from being cut by shooting up new growth and extending its roots that will then decompose, building a rich and well-aerated soil. Then the growth slows down, as the grass begins to flower and produce seed. At this stage it will be unpalatable to sheep (though not to goats, who will help themselves to the woody stems of grasses gone to seed). Perennial grasses, if cut back again and again, and fertilized with animal manure, will increase the organic matter in the soil and increase its productivity, given the requirements of sunlight and rain. The biodiversity of healthy grassland—a salad bar for grazing ungulates—with its greater ability to hold water, insures resilience through periods of both drought and flooding.

"A good solution solves more than one problem, and it does not make new problems," Wendell Berry famously said. Industrial agriculture, by removing animals from grass, has created two problems out of what was an elegant solution to both: we begin with the miracle of ungulates with their four stomachs turning grass, sunlight, and water into flesh and bone and sentient life through a cycle in which the animals nourish the soils that in turn nourish them—no need for chemical fertilizers, no two-acre manure lagoons required. Industrial agriculture, by removing animals from pastures, has broken this eternal link, the *permanence* of the animals in John Berger's poem, who once "flowed like their milk." By putting grazing animals back on grass, regenerative agriculture aims to restore the soil as the eternal link, reestablishing the permanence of farming practices that can endure for generations. A healthy soil, rich in organic matter, will hold more water, increase biodiversity, and reduce erosion and runoff, all while increasing the health and happiness of the animals—and the farmers who care for them—who are allowed to express their "characteristic form of life."

As Michael Pollan reports in *The Omnivore's Dilemma*, a well-managed pasture is many times more productive than a continuously grazed rangeland. It is also, Pollan argues, sucking heat-trapping gas out of the atmosphere and storing it safely in the ground. "If the sixteen million acres now being used to grow corn to feed cows in the United States became well-managed pasture, that would remove fourteen billion pounds of carbon from the atmosphere each year, the equivalent of taking four million cars off the road." A well-managed pasture can produce as much biomass as a forest, Pollan claims, sequestering as much carbon in the soil.

There is no doubt that a well-managed pasture has many benefits, but carbon sequestration will decline over time (what Pollan fails to mention in his book); once the soils are healthy, they are merely storing carbon, no longer absorbing it from the atmosphere. It is not clear at what point the grasslands will stop being a climate benefit (research estimates about thirty years), no longer offsetting the methane emissions from the sheep's digestive systems that continue once soils are stabilized. Pollan also might have calculated the climate benefit of not just putting all those feedlot, corn-fed cows onto grass but trading in those cows for chickens, say, or cultivating legumes to feed people rather than livestock. He omits any mention of the methane emissions from the digestive systems of ruminants (the world's 1.5 billion cattle alone emit a massive 231 billion pounds of methane annually, a more potent greenhouse gas than carbon dioxide), which subtract from any climate benefit to grazing animals on well managed pastures.

Pollan's book was still generating a buzz at a time when the "two percent solution" was gaining currency, a time when global efforts to reach a climate agreement were at a stalemate and the world was desperate for hope. Researchers had determined that a 2 percent increase in the organic content of the earth's soil could soak up *all the excess carbon dioxide in the atmosphere within a decade.* Some European countries took notice, and so did some billionaire ranchers in the United States, who set out to prove the viability of "carbon ranching" by creating pilot projects in regenerative grazing.

One is the Marine Carbon Project, a collaboration between California rancher John Wick and University of California, Berkeley researchers. Another is TomKat ranch, a regenerative ranch of eighteen hundred cattle owned by the billionaire climate activist and one-time presidential candidate Tom Steyers. It was an attractive idea: not only might we continue to eat beef but we might suck out all the excess carbon in the atmosphere while doing so. "Imagine what it would mean if a net-zero-emissions cattle farm were as big a symbol of American achievement in fighting the climate crisis as an electric vehicle," 2016 presidential candidate Pete Buttigieg enthused. But if "net zero" cattle ranching sounded too good to be true that is because it probably was.

The results of both experiments in sucking carbon from the atmosphere by grazing cattle turned out to be underwhelming. While evidence for the benefits of regenerative methods was unequivocal—increased biodiversity, reduced erosion, and healthier herds—soils on Steyers's ranch did not show increases in stored carbon, as *Politico* reported in a 2019 article. John Wick, too, would have to temper his claims for the promises of "carbon ranching," having found that his improved soils didn't store carbon as he had hoped, while soil carbon proved to be too difficult to measure with enough reliability to be traded in a carbon market. Neither Wick nor Steyers present any data on methane emissions in their analysis. Even if good management practices show increases in soil carbon, what does it mean if those gains are cancelled out by the methane emissions from their cattle?

It may be that as a solution, soil carbon sequestration came too late. A warming climate is making it harder for soils to do this job, which may account for these disappointing results. If a feedback loop is already underway in which warmer temperatures are speeding up the decay of organic matter in soils, thus releasing their stored carbon, then the need to protect carbon-rich soils becomes all the more urgent. But ultimately, as the MIT Climate portal concludes, "soil-based carbon sequestration, like other negative emissions technologies, cannot take carbon out of the atmosphere as fast as we are currently adding it. To stop global warming, these efforts to store carbon must be coupled with drastic cuts in greenhouse gas emissions."

Newer research does include methane emissions in studies of the climate benefits, or demerits, of grazing animals. A comprehensive 127-page global study from 2017 conducted by the Food Climate Research Network at the University of Oxford found that the climate value of grasslands has been overrated. Another 2023 study that compared emissions from the practice of transhumance—that is, a traditional form of mobile pastoralism in which sheep are moved between winter and summer pastures—to intensively raised, grain-fed sheep, concluded that mobile pastoralism could play a role in climate mitigation. However unsettled the research is, it does not come as a surprise to me that studies are finding traditional methods of mobile pastoralism to be healthier for the climate than industrial production. A true measurement must distinguish between them. And more importantly, cows, sheep, hogs, and poultry are not equal partners in their contributions of heat-trapping emissions, with beef cows responsible for an outsized share. Without inflating the benefits of "carbon farming," the potential to reduce the climate impacts from agriculture, which is responsible for up to one-third (estimates vary) of all greenhouse gas emissions, by deindustrializing farming, is immense—fourteen billion pounds a year of carbon saved, just from moving cattle out of U.S. feedlots and onto grass. That would be just the beginning.

For all those years, I, too, held to the promise of grass farming and its potential to sequester heat-trapping gases. I now see that it is all the trees we planted that are doing the work of storing carbon on our land—you could say, we are *silvopasturalists*, who have planted willows, maples, fruit trees, and white pines to create an open parklike landscape where our animals graze. Still, a true measure of the carbon balance of raising sheep should also consider the many services and benefits of these animals, who are not just providing meat but also wool, mowing services in the place of gas mowers, and what may be most valuable of all—manure for our gardens, that black gold. If we were to get the full efficiency of our animals, we would also be milking them and making cheese. The same grasslands that serve as sheep pasture are also sustaining chickens, and gift us with an abundance of apples, pears, and maple sap; loads of edible weeds and herbs; and kindling for our woodstove. They provide dandelion, clover, and goldenrod for honeybees; hunting grounds for hawks; and habitat for a multitude of pollinators and insectivorous and seed-eating birds—sparrows, swallows, finches, robins, mourning doves, chickadees, and blue jays being among the many feathered and winged creatures in a complex web of life.

It did seem to me to be very labor intensive, but the amount of time Art spends moving fencing amounts to a quarter of the time he might otherwise spend sitting on a lawn tractor. And besides, he gets from his efforts more than a pasture: food and wool, manure for the vegetable garden, exercise, and the company of a nonhuman family. That is all part of the equation, too. To choose this life is to affirm a culture shift, one that might solve the twin problems of industrialism—the problems of *waste*, and the insatiable need to *acquire*, bedazzled as we are by all things fast, electric, and new. I will put my faith in the scythe, the wool sweater, and the bicycle before the lithium-ion battery or the electric car. "The revolution won't happen by people staying home," argues Rebecca Solnit, who thinks that individual acts of thrift won't get us the distance. And yet this is how India defeated the British Empire—with a drop spindle. This is the meaning of *direct action*: we are not waiting on change but are *building the future now*. Maybe our noncooperation, our staying home, *is* the revolution. Maybe it is the most radical thing we can do.

Consider the Lambs

I was a practical vegetarian long before I was a principled one. In college, when cooking meat at home was just not something that we did, I learned to cook from the *Vegetarian Epicure* and *Moosewood*. It was only later, in my midtwenties, after reading *Diet for a Small Planet*, where I learned that 90 percent of the grain grown in the United States was fed to livestock, that I swore off eating meat altogether.

I began to make exceptions: Thanksgiving was to me the one time of year when, I felt, the proper ceremony and respect were given to the animal who was feeding us. And then, another exception, when I moved in with the man who would become my husband, who raised sheep, I began to eat his lamb.

It may seem odd that a vegetarian would move in with someone who raised and slaughtered animals. Or that, of all the meat I might enjoy eating, it would be lamb—which probably has to do with the cuisine I grew up eating, at the Middle Eastern restaurants near my childhood home. Each winter, I would dread the day when someone would come to the farm to "harvest" the one-year-old lambs. I stayed inside the house until it was over and would never see or hear a thing. After, the ewes would be going about their day as if nothing unusual had occurred. They did not even miss their lambs. This is altogether less stressful for them than shearing day—when the lambs and their mothers can't figure out who is who after their haircuts and for hours it is mayhem.

It was this peacefulness, on the day of the most violence the herd would see in a year, that informed my feeling about eating meat. We *were* a peaceable kingdom, which I now understood could admit the practice of slaughtering animals for food. Still, it was a sad day. "It's very difficult to deal with," my husband said to me once, who is not generally one to express his feelings. There would be something wrong if it were not, although he chose to do this, and to assist the farmer he hired to do the slaughter. He would hold the lamb's head in his hands when the gun was drawn. I know that the lambs did not suffer because he was right there to witness this and he knew, as much as we can ever know.

Nonetheless, in the first years I often thought that we should end the practice. But to put animals on the land to provide mowing services and wool, without also producing food, felt somewhat wrong to me. And it would mean no more lambs. No more lambing days. No more newborns to watch learn to stand and nurse, their little tails trembling. No racing or playing king-of-the-mountain. It is because people eat lamb that these animals have any life at all. Goat dairies will euthanize the male kids at birth, because there is no market for goat meat. (I will never get over reading Melissa Coleman's account, in her memoir *This Life is in Your Hands*, of her father, the organic farming trailblazer Eliot Coleman, a Nearing protégé, drowning a newborn goat kid in a bucket of ice water because it had the misfortune of being born a male.) Is it better that these lambs were never born, or were drowned in a bucket at birth ("Better to get it over with," Coleman said), than to have "one bad day" in their otherwise happy, short lives? Even Peter Singer, the ethicist who wrote *Animal Liberation*, agrees that an animal given a good life on a farm like ours, where the animals are allowed to express their "characteristic form of life," are better to have lived and died than not to have lived at all.

The farmer who slaughters lambs or calves is known to wail and grieve for days. They are not without feeling. Art does not wail out loud, but we are both, in the aftermath, especially quiet. Our sensitivities are heightened, as if we had peeled away a protective layer around our hearts and exposed them to the world's primordial pain. We chose this life not because we are callous but because we want to feel more deeply than we will by living in urban detachment from the realities, and the tragedy, of existence. We crave more reality, more feeling, not less.

The choice not to eat anything with a face, or anything that crawls, or wiggles, or creeps, or only fruits and vegetables that fall from a plant—these are spiritual, intimate, and cultural choices that reflect just how much violence, or the kinds of violence, we can spiritually or emotionally stomach, because all food production involves violence. ("Don't let anyone tell you that growing vegetables is not a violent act. The muted sound of a plow tearing through roots is almost obscene, like the sound of a fist meeting flesh," writes vegetable grower Kristin Kimball in her memoir *The Dirty Life*.) In a confusing "melting pot" culture with too many choices, to control what we eat can be the beginning of taking control over our lives. And to choose vegetarianism or veganism is often a first step in considering eating as an ethical, and agricultural, act.

After all, people have respected my vegetarianism for many years. While I respect the choice to be vegan, I take issue with those arguments that fail to distinguish between different kinds of farming, with their different standards of care, or that misunderstand the nature of the interspecies symbiosis that is domestication. Those videos on PETA sites are not images of what goes on at small farms, where farmers are surely not in it for the profit but for the love of it—and for the love of animals, with strong ethical convictions about environmental stewardship, and care and respect for animals. Herbicides or plastic weed barriers or tilling? Allow a calf to nurse and therefore lose efficiency, meaning not only loss of profit but a greater climate impact? These are some of the choices a farmer must make, and all farming practices involve trade-offs because to live at all is to be subject to the laws of entropy and all life is dependent on death.

"For years I'd taken imagery on dairy-milk cartons literally, peaceful cows standing in fields beside gentle farmers seated on stools, red barn in the background under a vast open sky," writes Lisa Monroy in her essay "Soli/dairy/ty," an account of her conversion to adamant veganism after developing an empathy for the lactating dairy cow, as one lactating mammal to another. Pregnancy and birth for a cow, she finds out, are "almost identical" to those of humans, and when she watches videos of a nursing calf, she says, "That is exactly like Alexandra!" (her own baby). Exactly. "Was that the real propaganda? In YouTube videos of the routine dairy-farm practice of taking newborn calves from their mothers, the distress cries sound chillingly like daycare drop-off, except the afternoon reunion will never come."

It is true that even on Old MacDonald's farm (if it's a dairy farm), the calf, kid, or lamb is most often taken from its mother soon after birth. Even the gentle farmer on his milking stool, who *does* have a red barn and whose animals do live mostly under a blue sky, will practice this "subterfuge" as the goat farmer Brad Kessler describes it in *Goat Song*, a memoir about his first year raising goats: "The happy calf sucking its dam may be our storybook image of pastoralism, but it almost never happens in a commercial operation. Every dairy we knew—cow or goat—pulled its calves or kids from their mothers right away or a few days after birth."

This is a farmer who writes poems about his goats, who calls his milk stool his *prayer mat*, and who anguishes over a milk goat when she falls ill. "We decided to do the same with our kids. We'd milk the mothers and dole the liquid to the newborns. In this way we'd imprint ourselves on the kids and they'd be tame and friendly and not afraid of humans. We'd be all things to all the goats: mother, kids, milker, herd king and queen."

He has made up for the violence (his word) of this forced separation in spades with tenderness and care.

It is true that as mammals we have much in common with the dairy cow or goat. To recognize our kinship with other creatures is the beginning of empathy toward them. But are we *exactly* the same? I have spent some time among dairy cows, I have seen the calves taken from their mothers soon after birth, and I have spent many years around ewes and their lambs. We *beg* our sheep to raise their own, but when they won't, or the lambs can't nurse, their mothers show no distress when we scoop up their newborns to raise them on the bottle. I cannot imagine a woman trampling over her dead baby while eating her breakfast. I cannot imagine her simply walking away when her baby needs to nurse. I cannot imagine a woman tossing her newborn against a wall, as I have seen a ewe do for no good reason that I could fathom. Neither will a woman get down on all fours to slurp down her afterbirth, as if it were a giant bloodied oyster, as cows will do (but, thankfully, not sheep).

"The birth is identical." But it is not. Anyone who has witnessed the birth of a lamb or calf cannot help but to be impressed with how easy it all seems. Our babies are born prematurely, and need that close bond with a mother for much longer than a calf or lamb, who is on her feet in no time, and after a few weeks no longer even needs her mother's milk. An orphaned lamb who is bottle-fed *does* need a herd but does *not* need a mother ewe at all—unlike a human baby. Our bottle lambs live with the herd just as the other lambs— curling up with their cousins, brothers and sisters, playing with them, following the adults to new pastures. This is not to say that it is not harsh or even cruel to take the babies away—it is to say that cows and human mothers are *not* the same, and their differences, from an ethical point of view, may be crucial.

Kristin Kimball, one of those gentle farmers who sits on a milking stool and milks by hand, whose animals do spend most of their time under a blue sky, in her book *The Dirty Life*, offers this description of the separation of a newborn calf from its mother, Delia: "Mark wrapped a towel around the heifer, hoisted her onto his shoulders, front legs in his left hand, hind legs in his right, and walked toward the barn. We expected Delia to follow but she couldn't understand where her calf had gone. She nosed the grass where she had given birth, wondering if it was still there but invisible. She bawled urgently and would not move from the spot."

Then the other cows headed outside and Delia followed. "As soon as she was moving, Delia stopped bawling for her calf. From then on, it was as though there had never been a calf, or she had willed herself to forget."

We cannot know what an animal is feeling because our communication with her is limited. But when a cow wails at first, when her calf is taken, as she searches the ground for it, I think it is safe to say she is distressed—but if we can say this, can we not equally say that, one hour later, when she seems to have forgotten all about her calf, that she has recovered?

In truth, it is our kinship with cows and goats that allows this relationship of mutualism: that we can communicate—sharing certain signals across the species barrier to signify submission, acceptance, pain, and affection. Sheep, goats, cattle, and horses are all herding species with enough in common with us that they are able to accept us into their social hierarchies as group leaders. Our sheep follow us because they accept us as members of their group, not because we hold them captive, or because we force them to follow at the end of a whip. It is why, when a farmer reaches her hands under a goat's udder to milk her, the she-goat accepts this hand on her teats as much as she would the mouth and teeth of her own kid.

Not every dairy will separate mother and newborn. The goat dairy where I worked in France did not do this. By the time I arrived in early summer, the kids had already been weaned. When I led the troupe out in the field in the mornings, and again in early evenings, kids, mothers, and billy goats were all together as a single herd, the kids rambling together, knocking their horns together in play, running alongside the adults. For this family, cheesemaking was a seasonal practice, as it is for traditional artisanal cheesemakers in many parts of Europe. In the Roquefort region, farmers allow their lambs to suckle for three months; so do the Basque cheesemakers, who believe it is important to "respect the seasonal nature of sheep." (Were we to milk our sheep, we would certainly allow the lambs to nurse.) Some cow dairies are experimenting with this here in the United States but the economics of dairying here make this next to impossible for any farmer who hopes to at least break even.

By failing to distinguish between farms with different standards of care, Monroy is hurting small farmers who are providing an alternative to the industrial feedlot, its "planet damage," and its "bull masturbators." Monroy writes, "I wonder if, by attempting to raise two vegans in this era of awareness and better substitutions, the domino effect can spread to the point where future generations will look back on an uncivilized past society of bull masturbators and planet-damage just so people could engage in the bizarre custom of drinking another mammal's milk and think, Why did they ever do that when we get a similar-enough substance from nuts not attached to sentient beings? A mother can hope."

Why did they ever do this? is to me a question worth investigating. I am not so willing to skip over, as Monroy does, not only most of human history but also how people around the world currently live, who may not have equal access to the wonders of historical progress (those "better substitutions") and never will. For many people, especially women, a dairy goat or cow can mean the difference between life and starvation; a small flock, between poverty and a good life. A globalized food system that gives us the option of buying imported, locally inappropriate foods shipped over many miles may not be a better substitution for everyone, and in fact that global food system may be the reason for the impoverishment of billions of people. If sheep, goats, and cows have been so successfully adopted by people in almost every region of the world except the Arctic—and in places where other forms of agriculture (growing almonds, for instance) are not viable—over many thousands of years, there must be reasons why humans adopted this practice instead of becoming universal consumers of almond milk.

"From the cow comes a whole farmstead of abundance," writes Kimball. "Milk, cheese, butter, yogurt, cream, and the by-products— skim milk, buttermilk, whey. She is the cornerstone of the farm."

Why did they ever do that?

"You know, o my son, the first thing God created when he created the world was milk. And from their milk came life. . . . Those who follow the camel and live most exclusively off their sweet milk are the ones who live in greatest harmony with the universe." Were Monroy to ask her question, *Why did they ever engage in this bizarre custom?* of this Bedouin of the Empty Quarter, Al Hadayb, or any of the herders of the world, going back in time since nomadic pastoralists first realized they could live from their animals without killing them, the answer would sound something like this: because milk is life.

And because a milking goat, sheep, or cow can give us everything we need in order to live, it also means freedom. Goatwalker Jim Corbett writes, "Archeologists generally maintain that, with a few possible exceptions, such as Lapp nomadism, pastoralism emerged as a spinoff from cultivation. . . . Some human beings, after learning how to earn their bread by the sweat of their brows, realized that a symbiotic relationship with ruminants opens an unguarded back gate to Eden; they simply took their animals and went feral." One does not have the option, once a regime becomes too oppressive, to pick up one's almond trees and simply walk away.

In truth we know very little about the origins of animal domestication. Humans adapted agriculture slowly, over thousands of years. That is to say, reluctantly and without enthusiasm. This transition was not so much a "revolution" as it was a slow erosion of the ecological conditions of existence that made any other choice impossible. Stephen Budiansky, in his book *The Covenant of the Wild*, writes that domestication was not an "invention" of human beings but was, rather, an evolutionary case of biological mutualism. Dogs and cats "chose" us—who would hang around our campsites, scavenging our middens and benefitting from the protection from predators we provided—Budiansky writes, rather than the other way around. Cats and dogs continue to choose to live among us, often hanging around when they are not wanted. As forests encroached on tundra and grassland in the retreat of the glaciers, grass-eating ungulates might have followed our clearings, and at any rate found themselves in close proximity with us, sharing our need for edges and savannahs. By hitching their destiny with ours, these species would not only survive extinction but would come to dominate. Budiansky writes, "A handful of minor species emerged from scraping together a marginal living at the end of the Ice Age to occupy a position of overwhelming dominance in the biosphere." Settled populations grew (not because cultivation produces more food but because nomadism is the best form of birth control), hunted out local populations of wild animals, and eventually came to eliminate the habitat for wild species. Both for the animals and for us, there was no turning back.

Helen Nearing and Scott Nearing, the elders of the back-to-the-land movement, which helped spawn a culture of vegetarianism, were ethically opposed to eating meat, although their position on dairy was a bit more moderate. (They did eat some cheese and yogurt.) "Carnivorism," the Nearings wrote in *The Good Life*, involves "holding animals in bondage," and "turning them into machines for breeding and milking." I don't believe our sheep live in *bondage* or in captivity. Yes, there are fences, but these are more like guidelines to them because they will breach them whenever they chose to. (Most of the time the electric netting is not even turned on.) Our sheep have no desire to take off—this is their home.

When I herded goats in France one summer, I followed a routine: I milked them in the early morning, then took them out for a walk until midday, when we all took a long siesta, and in early evening, I took them out again. One day, my host took me into Nivers, a medieval city nearby, to show me the sites. He was so taken away with his role as tour guide that he did not want to leave when it was time to take the goats out for their evening walk. I was anxious about keeping my appointment with the goats but he assured me it would be fine. *Ça va, ça va aller.* When we did get back, very late, we found the goats had busted down the barn door and were gone.

This was an illuminating moment for me, when I understood that domestication was a covenant we had with animals—that they only went along with our designs so long as they agreed to. Sheep and goats, in my experience, are creatures of habit, who lead very structured lives, and they don't like it when, without notice, we change the rules on them. That is to break the covenant.

It's still true that the brutality the Nearings saw in carnivorism is an ever-present danger inherent in the nature of raising animals for food, and is present in the massive feedlots where animals *are* treated like machines. And yet, farmers have been able to care about their animals *and* slaughter them. John Berger calls this paradox our "first existential dualism": animals were "both subjected and worshipped, bred and sacrificed." From the earliest records of pastoralism, we see the expression of the sacred nature of the relationship of the shepherd with his flock, with the good shepherd and the sacrificial lambs becoming central religious metaphors because of their importance as sources of life, and because of the spiritual nature of a daily engagement with the nonhuman Other. Farmers still feel this: Kessler writes that his relationship with his goats and with cheesemaking is "intimate and holy." "Pastoral metaphors of the nomad religions," writes Corbett, "about good shepherds who lead their flocks through desert dangers to cool waters and green pastures, who protect their herds from wolves and make every effort to recover every stray lamb—express the religious inspiration of people who regularly slaughter and eat the innocent lambs in their care."

To this vegetarian (of a sort), though I do not need to eat meat, I cannot so easily pass over thousands of years of human history—over our past as hunters who lived primarily on meat, or, since the end of the Pleistocene, the pastoralists who gave us our prophetic religions, our songs and much of our poetry. I do think that this has to mean something. Krishna, the Sumerian shepherd-god Dummazi, and Jesus of Nazareth were all shepherds, as were the prophets Moses, David, and Muhammad. What is it about herding, Kessler asks, and not, say, shoemaking or selling dry goods, that God chooses to speak to the shepherd? The herder who moves from the valley to the high mountains every summer—to the Zagros, the Alps, the Himalayas, the Hindu Kush, the Pyrenees—and back again for winter, wanders and roams under an open sky, exposed to that same whirlwind out of which God's voice came to Job. The religious texts of the prophetic age are rich in pastoral images because shepherding was a way of living in unalienated relation with the sacred gift of life and the mystery of death. The shepherds, too, have given us poetry and song, including the most beautiful poetry of the Old Testament ("The Lord is my shepherd"). From the songs the herders sang to their flocks came the musical and literary pastorals, and from there, a line extends through the poems of Hesiod and Virgil, to the English Romantics, through to contemporary rap verse.

Why did they ever do this?

Another question might be, Who would we be if they had not?

Lest we romanticize the pastoral age too much, as Kessler perhaps does, we need remember that these herders brought ecological devastation to much of the Mediterranean, as reflected in the epic of *Gilgamesh* and in biblical stories of the loss of Eden and the great flood. Still, for all its dangers, I don't see how a local food system in our region can do without livestock. (An alternative is to use green manures, that is, cover crops, but this requires a tractor and more tilling.) A mixed approach to farming, integrating animals and plants, is the basis of an ecological approach, as Sir Alfred Howard wrote in his *Agricultural Testament*, the foundational text of the modern organic farming movement: "Mother Earth never attempts to farm without livestock; she always raises mixed crops; great pains are taken to preserve the soil and to prevent erosion; the mixed vegetable and animal wastes are converted into humus; there is no waste." As Wendell Berry says about this passage, "That's it. That's the pinch in the hourglass."

I don't know how we would raise vegetables organically without animal manure, without rotating vegetable crops with pasture, and I don't know how a family would get through a New England winter without some dairy or meat in the larder. It would take a whole lot of beets and potatoes to make up for the nutrition provided by single lamb share or hog. The small farm with its flock of sheep and goats, its one cow, its free-ranging chickens, is the basis of a self-reliant, sovereign community, not wholly dependent on a fossil-fuel-based, technologically overdeveloped, and unsustainable global system. This is what the future of farming looks like, our postindustrial future.

In this vision, that bucolic farm with its red barn, old maple trees, and pastures dotted with sheep and cows will once again take a central place in the agricultural landscape, with the consumption of meat and dairy in our diets looking a lot more like it did in the ancient world—and like the diets of billions of people in the world today. This is an image of a culture that has learned to eat mostly plants, with some meat and dairy, but not much. The World Resource Institute estimates that a growing population with its growing appetite for meat will drive the deforestation of a land mass twice the size of India. This the world cannot sustain. The higher cost of meat and cheese will mean we will eat less of it, and will be offset by the lower cost of eating more beans and nuts, and more eggs and poultry in place of beef. Farms that integrate livestock with horticulture and agroforestry, that rotate cows, chickens, and pigs on shared pastures while also supporting honeybees and wildlife, are *more* productive than factory farms that produce only one thing (besides pollution). Those plastic-wrapped hay bales, for all their efficiency—responsible for an annual 2,700 tons of unrecyclable plastic waste in Vermont alone (one ton is the equivalent of thirty-six thousand sixteen-ounce plastic bottles)—will also be a thing of the past. Why did they ever do this? A person can hope.

The farmed landscape of the Lake District in England is a basin of enclosed land—a patchwork of farmsteads and meadows, criss-crossed with stone walls and hedgerows—surrounded by uplands locally called *fells*. In summer the shepherds move their ancient breeds of Herdwick and Swaledale sheep from the lowlands into the fells to graze the common lands, where the herds drink from streams and shelter from wind and rain beneath the big Scots pines. The summer meadows, allowed to flower and go to seed, are abundant in this lush region of heavy rainfall, passing from pale green to yellow to russet as the spring passes into late summer and then autumn, before they are mowed for hay for the winter, when the shepherds will gather their flocks from the uplands with the help of sheepdogs, calling out to them in their distinctive songs, working in concert with the other herders with a common purpose—dogs, herders, sheep—as they have done for generations. A rhythm of seasonal life exacts an exchange between the wild and the domestic, the fallow and the cultivated, with the bottomland hedged, drained, mowed, and tilled, and the rocky fells windswept and wild. It is the oldest farming system, and the largest surviving tract of common land, in Western Europe, and it has survived because the shepherds have continued the old ways. The people are "hefted" to the land just as the sheep are, whose knowledge of their place in the mountains was taught to them by their mothers as their mothers were taught in return.

James Rebanks, a shepherd of the district who described this life in rapturous prose in his memoir *A Shepherd's Life*, says, "There is no beginning and there is no end.... The grass comes and goes with the warmth of the sun. The farms and the flocks endure, bigger than the life of a single person. We are born, live our working lives, and die, passing like the oak leaves that blow across our land in winter.... Our farming way of life has roots deeper than five thousand years.... I love this place. For me it is the beginning and end of everything."

May it continue another five thousand years, with no beginning or end. I can think of no better substitution for such a life.

The Wild and the Domestic

Introducing Soays

Years ago, Art and I visited a place in Gros Morne National Park in Newfoundland called Green Gardens, a spectacular stretch of volcanic coast with stupendous coal-dark cliffs crowned with lush, velvet-green meadows and wind-twisted spruce trees. We camped for two days under a half-moon that lit the seafoam below as it swirled in and out of sea caves and skirted tall stacks of lava pillows.

The cliff-top meadows were once summer pasture for sheep, who still grazed there, though the farms and shepherds were centuries gone. It was a sheep's paradise. They would select the high points with the best views, we noticed, and bed down at night under the giant white pines. The scotch thistle flowers were in bloom, forming pink banners rippling in the ocean breeze. Lovely, except that the thistle, a plant that the sheep won't eat, was therefore taking over. The sheep were in effect selectively grazing themselves out of a habitat.

That was in the summer of 2004, when we weren't yet worried about thistle taking over our farm, as we were fourteen years later. Green Gardens served as a warning. Little by little, another variety of thistle, introduced here when the monitor barn was removed, was becoming a problem. This kind does not even have a pretty pink flower but rears its thorny head like a pair of ugly fangs. Everywhere. Especially around the footprint of the former barn. It was proving to be inextinguishable. We even saw it grow up through a great big pile of wood chips, as if it were clawing its way straight out of hell.

And then another invasive grass that the sheep won't eat—we call it *jaundice grass* for its sickly yellow hue—was becoming even scarier.

We considered introducing a pair of goats. But that would involve some serious fencing if we hoped to have a tree left standing. I shuddered to think what they could do to all the hard work of the farm crews and volunteers next door. Not to mention our laundry.

Then we learned about Soay sheep.

Soays are a rare breed of primitive sheep that were isolated for two thousand years on an island in the Hebrides. They are thought to be close to the earliest domestic sheep herded by Bronze Age shepherds. They are small, and horned, and resemble deer rather than ovines. Having lived on an island not unlike the cliff-top meadows of Newfoundland, they would have learned not to be such fussy eaters, once there was nothing left for them to eat but thistle. Soays do not need shearing, for they shed their fleeces naturally, nor nail trimming, and even the rams are gentle, unaggressive creatures, which means they don't need to be separated from the flock. They are a good choice for farmers who are reluctant—as we are—to intervene too much in the lives of our animals but rather wish to live with them in a spirit of mutualism. You give me wool. I give you water.

The day came to pick up a pair of Soay wether lambs from the Soli Deo Gloria farm in Whiting. We caught our first glimpse of them in the field—I might have mistaken them for deer except for the way they were so tightly flocked together. The farmer led them into a corral, and as he tried to grab one for us, they moved swiftly, swirling around him in a blur like a school of fish around a diver. "They're impossible to catch," he said. As we were soon to learn.

We brought them home in a U-Haul minitrailer and let them out in the paddock to mingle with the other sheep. They retreated to a hidden corner of the paddock behind the barn, where I stood and watched them for a while, and they, in their shyness, returned my gaze. These are animals who will remain calm in your presence, and seem to be as curious about you as you are about them, as long as you move slowly and make no hint of giving chase. Their curved horns sweep back from their faces, marked with white swooshes of war paint, giving them a noble look. I was excited to observe them over the next few days to see how they would adapt to their new home.

But the next morning, Art came into the house from the barn, saying, "They're gone." I got dressed and followed Art down the road and into a thicket of poison parsnip running into a ravine, where the lambs were last seen. We walked up a little streambed into the woods and came out at the edge of the VYCC farm. No sign of them.

Paul Feenan, the farm program director, was there with his youth crew, giving the teenagers instructions on their morning's task. Never missing an opportunity for a teaching moment, he said, "Let's introduce Art and Alexis."

He explained that we are the ones who live next door, with the sheep. "They help us out sometimes, and we help them out at times. Right now is one of those times."

We explained our situation, and described the lambs to them. "They look like tiny antelopes," I said.

"What's an antelope?" one of them asked.

As it turned out, we were very lucky that the crew happened to be working there that day and would keep an eye out for our lost lambs.

We bushwhacked through the jungle again, but still no sign of them—only a few of their tiny triangular prints in the dirt where they entered the ravine. The understory was dense with tall weeds, and the woods merged into the town forest and continued on toward Canada. We despaired that we would ever find them, and even if we did, as the farmer said, "They are impossible to catch."

Later that day, I received a phone call from Cae at the VYCC. "Our crew has spotted your lambs. They were seen poking their heads out of the woods. Brown. With horns."

I ran back and there they were, out in the open, noses to the ground like a pair of brown dogs sniffing around. I followed them for a while, slowly, as they made their way down the path at the edge of the woods and then meandered into the tall weeds, where they munched happily on leaves from the low-hanging bough of a box elder. They looked at me, and we exchanged gazes again, and they went on grazing. Like Jane Goodall and her chimps, they accepted my presence as I lingered and observed.

I had no way to get them home, of course, but it was a source of hope that they had stayed nearby, and did not take off into the deep wilderness, as we had feared. When Art returned from work, we went back to the same spot and found them again, their war-painted faces visible through a slight gap in the weeds, like an image from an Henri Rousseau painting. It seemed fitting that these creatures, at the confluence of the wild and the domestic, should find themselves on the line between forest and farm. They would have everything they could ever want here, I thought—water, shade trees, and good browsing. Everything but a herd.

They were, to our dismay, choosing to bed down in the exact place where we leave gifts for the coyotes. However unlikely our success, we would have to try to grab them. Our plan was for one of us to come around from behind and flush them out of the weeds where Art and Cody—a neighbor who came to help us—would grab them. "Their horns make good handles," the farmer had said. Though we should know better than to try to chase sheep, who will always run from you. Far better to lure them to you with treats.

I crashed through the tall goldenrod and then crept up the streambed and climbed a bank to reach them. To my surprise, they did not notice me until I was standing over them. They looked up at me with those sweet faces, and before I could apologize—they bolted. I grabbed one—but the other got away. He went bounding past Cody and Art up the path toward the lean-tos, graceful as a deer, and then leapt into the woods and vanished.

We looked for him for two days. I got to know the edge of the woods, the little streambed, and the new farm crew and apprentices, who were by now all invested in helping us to find our lost lamb. I could not help but think about him, scared and alone, and possibly lost, in the deep woods. As I focused on my tasks—in the garden, in the kitchen, or at my desk—those woods were at the periphery of my consciousness. The endless possibilities—some hopeful, some horrific—turned over and over in my mind. Nature, I know, has its cruelties, but we were responsible for this lamb, who we had wrenched from his family, grabbed by the horns, and brought to this strange place with its gang of bully-ewes. In truth, I do not like animal husbandry with all its little brutalities—the rassling, the chasing, the penning in. And so much death. There are times when I wonder why on earth we are doing this: this was one of those times.

Meanwhile the lamb we brought back home settled in. It took only a day for the other sheep to accept him—they were mean to him at first, which is probably why the Soays ran away—the new lamb eager to become a part of the group. Seeing how closely he stayed with the herd, I was hopeful that the other lamb would find his way back to his brother.

He did. After two nights alone in the woods, one of them in the pouring rain, he made his way across the open farmland under the cover of a morning fog. We opened the gate for him, and watched as the settled lamb led him proudly, like a college senior showing the way to a first-year student, into the barn.

Now the two of them are together again, inseparable, and show no inclination of running away. (We named them Lost and Found.) These gentle creatures are teaching us to be gentle ourselves, who will never again give them chase, and should never have reason to use their horns as handles, as they live out their lives here in peace.

La Dignidad

While we grow our own food as an exercise in food sovereignty, I have not lost sight of the truth that we still live in a global economy that deprives others of that sovereignty. I have been given the gift of land where I can grow my own food and exercise personal freedom. Meanwhile, not far from where we live, and on large dairy farms across the state, migrant farm workers are living and working without basic rights, precisely because they have been deprived of their sovereignty by economic policies to which U.S. lawmakers have subscribed.

When news about the presence of migrant workers on Vermont dairies first came to me in the late nineties, images of the *Grapes of Wrath*, Cesar Chavez, and the grape boycott came to mind—large corporate farms worked by exploited migrant workers—that seemed incongruous for a place where families still owned and worked their farms. Within a decade, Vermont dairies would become almost entirely dependent on undocumented migrant workers, who do not qualify for H2A guest worker visas issued for seasonal agricultural work (what migrant justice advocates consider, anyway, to be a system of "legalized human trafficking"). As Vermont Governor Peter Shumlin would say in 2011, "Vermont farms can't survive without workers from outside America. It is just the way it is." Following Trump's election in 2016, Vermont's Agency of Agriculture developed a plan to deploy prison labor to Vermont dairies should the undocumented workers be deported, putting to rest any doubts that these workers were stealing good jobs from Americans willing to do this work.

In the early aughts, I joined a group of volunteers who began to
visit these farms to offer English classes to Spanish-speaking migrant
workers. The workers expressed surprise that we had come. *No one
came to us in Florida,* they said, *where we picked strawberries.* Later
the desire to support migrant workers was channeled more usefully
into Migrant Justice, a worker-led rights organization that, over the
next two decades, would secure undocumented workers the right
to obtain drivers licenses; advocate for migrants detained by ICE;
campaign for fair and impartial policing policies; secure stimulus
payments for migrants during COVID; and launch an ambitious
long-term Milk with Dignity campaign for better working conditions and fair pay.

That was also the time (the late 1990s and early aughts) when I
worked as an advocate for Rural Vermont, a small farm advocacy
organization born in the farm crisis of the 1980s when Vermont's
prosperous dairy industry came crashing down in one farm bankruptcy after another. At the time, when the first two-thousand-cow
dairy barns were under construction, Rural Vermont saw that the
move toward getting bigger—toward greater industrialization, more
debt, and greater corporate and technological dependency—would
not be good for small farmers or for farm communities. We opposed
GMOs. A member of the international Via Campesina, the National
Family Farm Coalition, and the Farmers Union, Rural Vermont
promoted a supply management system (meaning, farmers could
get off the treadmill of producing ever larger quantities of milk)
that would stabilize prices and eliminate the need for subsidies,
and we opposed the globalization of food systems.

We are more fortunate in Vermont than in most of rural America,
which has suffered under these developments. Our small towns
are vibrant, and the storefronts are not shuttered as they are in
small towns across the country. Agriculture has survived, but it has
changed here over the last twenty years. On one end, a local food
movement is thriving. On the other, industrial dairy production
has been concentrated into two watersheds, and only a few large
farm operations, with consequences not only for water quality. The

stench of a CAFO is not the quaint smell of yesterday. Farmers were pressured to get big or get out—by banks, agribusiness corporations, and extension agencies—and were compelled to increase production to make up for low prices, which only depress prices further, and thus they have been on a never-ending treadmill. The bucolic image of the Vermont landscape dotted with cows grazing on hillside pastures, with red barns and silver silos, has given way to the surreal landscape of multi-million-dollar barns the size of airplane hangars, plastic-wrapped hay bales that look like giant marshmallows, and landscapes not only without cows but without people, where undocumented workers hide in the shadows. Aldo Leopold described it, it's the difference between the farm as a "place to live" and the farm as "food factory."

Meanwhile, the Vermont dairy farm continues in its slide toward extinction. From 2020 to 2023 Vermont lost 128 dairy farms, but the amount of milk the state produced remained the same, at a staggering 2.5 billion pounds. This means farms are fewer but getting bigger, and so are the pollution problems for Vermont's waterways. And in that same time period, organic producers also declined in number, as prices for organic milk, which had previously enjoyed a premium, slumped below the cost of production.

It is the migrant farmworkers who are keeping these farms from tipping over the brink.

More than half of the 2.5 million seasonal workers on U.S. farms are undocumented, with fear of deportation acting to depress wages and suppress any movement to challenge abuses or demand better working conditions. The precarious status of farmworkers in the United States, who are primarily people of color, is owed to their exemption from minimum wage requirements and other basic labor rights; this is the legacy of the racism built into New Deal programs—FDR's concession to southern lawmakers who controlled key congressional committees. "One of the goals of immigration restrictions throughout U.S. history has been to maintain an under-class of workers without rights protections who can be exploited and abused," says a Migrant Justice activist. "Now is no different."

So far Ben & Jerry's is the only milk processor to have signed a Milk with Dignity agreement with migrant farm workers, a legal and enforceable agreement that ensures their basic rights are met. In a state that prides itself on its historic opposition to slavery, almost half of Vermont migrant dairy workers—those who do not get the benefits of the agreement—report they don't get a rest day, nor do they get paid a minimum wage. Their housing is substandard and overcrowded; many have their first paycheck withheld; and they are often not paid on time. They get no health care, no childcare, no sick or family leave. Worst of all, they are *encerrado*—confined to the farms for fear of being seen and reported to immigration, as they often are.

"This is the first year that I can take a vacation. It feels good to get a breather. We feel more secure, safer now that we have the right to speak up, and to not be fired," one worker testified about the benefits of Milk with Dignity.

"I feel more secure knowing my rights and having all these benefits," said another. "Before nobody cared if we got sick. We had to work and if we couldn't, that day was taken out of our paycheck. Now that we have Milk with Dignity, we're paid that day. And that's important for me, for all of us."

No other milk processor has stepped up and volunteered to join the Milk with Dignity program, which requires processors to pay farmers a premium for their compliance. The supermarket chain Hannaford, which has been the subject of a sustained Migrant Justice campaign since 2020, has dug in its heels, even as they are the target of a boycott.

One unforeseen consequence of getting big that I did not foresee is that a two-thousand-cow dairy operation requires round-the-clock labor—dirty, dangerous, and demanding work, involving ten to twelve or more hours a day bent over a cow's rump. Over my first winter living in Vermont, in 1993, I worked mornings in an industrial dairy barn (small by today's industry standards) that was adjacent to the house where I rented a room. The ten dollars a day I was paid for helping with morning chores did pay my rent, but mostly I chose to do this because I was curious about *work*, and about farm work in particular. I liked the time I spent with the farmer, who was grateful, above all, to have some company in the barn, where he otherwise worked alone in the hours before dawn, and I would not regret that I had done this. But I did not like the work, and I could not have imagined doing it for ten or twelve hours a day, every day, year after year. I especially did not like the stench; I did not like getting swatted in the face by a manure-laced tail; I did not like working in the dark, pressed up against cows chained to their stalls twenty-four hours a day until spring, when they were put out to pasture. Working there was not an existential necessity for me but a choice, of which I had many, because choice was my birthright as a white U.S. citizen with a passport, a social security number, and a college education. Unlike some of the migrant farmworkers today, I also had private bathroom where I could take a hot shower to wash away the stench.

"Small farms can't compete," is the common explanation for the failure of small farms, as if farmers were in a Darwinian competition with one another, may the strongest one win. ("It is just the way it is.") The decline of the family farm, according to this narrative, is the inevitable price of modernization, rather than the work of policy makers who set out to deliberately dismantle New Deal Programs in favor of free market policies that allowed farm prices to go into free fall. Meanwhile, agribusiness middlemen, petrochemical companies, and banks all benefit from larger, more technologically and chemically dependent farms. *Fewer farmers* was explicitly the goal, as it would later be in Mexico under NAFTA. "American agriculture's like a big pie," said Earl Butz, USDA secretary under Nixon. "Right now we've got lots of farmers, and each one is getting a small slice of the pie. We need to eliminate a bunch of them, so that those that are left will get a lot bigger slice." Growing inequality, in other words, was not an accident but a goal.

It is not a coincidence that migrant farmers started arriving in Vermont soon after NAFTA went into effect on January 1, 1994.

(Promoters of NAFTA had promised it would stem the tide of immigration). The architects of NAFTA on the Mexican side used the new economic framework to engineer the end of small-scale farming and traditional food systems, just as policymakers had done decades earlier in the United States ("we need to eliminate a bunch of them"), by flooding the market with cheap U.S. corn, in the form of processed foods and animal feed. Some two million small farmers and farmworkers would lose their livelihoods in the next decade, and many of those farmers would migrate to the United States. Over the next decade and a half, half a million people *per year* would emigrate, with 10 percent of the Mexican population living in the United States by the year 2006.

The destruction of the traditional *milpa*-based food system in Mexico would not only come at a time when gourmet Mexican cuisine would be championed by celebrity chefs and fine restaurants in the United States but precisely at a time when the world was looking for models of sustainable agriculture—practices that could thrive without irrigation, fossil fuels, or chemical "inputs." And yet the gold standard of sustainable farming—where plant varieties, cultivated over generations, were resistant to drought and to *plaga*; where corn, beans, and squash could be grown year after year, without exhausting soils—has been all but pushed out by monoculture farms and U.S.-style animal feedlots, dependent on petrochemicals, GMOs, and more debt.

It is a fact that farmworkers, Vermont dairy workers among them—who work long hours, have little time off, and poor access to healthful or culturally appropriate foods—are among the most food insecure in the country. Allies of farmworkers, in addition to supporting the Milk with Dignity campaign, can volunteer with the Huertas Project, a University of Vermont Extension program that assists migrants in creating kitchen gardens that can restore some personal food sovereignty to workers who have little choice but to sustain themselves by foods with little nutritional or cultural value. With the assistance of interns, seeds, and plant starts provided by Huertas, farm workers grow beans, squash, tomatoes, cilantro, calabaza, and herbs in small gardens."Huertas allows the possibility for farmworkers to cultivate food with deep cultural meaning, and to exercise agency and choice over what foods they consume," writes Teresa Mares in *Life on the Other Border: Farmworkers and Food Justice in Vermont.* "This is particularly significant as it is the denial of food sovereignty, specifically the dispossession of rural lands and livelihoods in Latin America, that has motivated so many farmworkers to move north."

Neither is it a coincidence that the rising wave of anti-immigrant sentiment has come at a time when extreme and disruptive weather events are wreaking havoc on communities throughout the continent, from sea to rising sea. As the ground shifts beneath us, one response will be to cling to our posts, to entrench ourselves behind Fortress America, holding on to an ever more rigid nationalist identity, an abstraction that is unaffected by the laws of physics or unruly weather. In his 2018 book *Down to Earth: Politics in the New Climactic Regime,* the philosopher Bruno Latour foresaw all of this, and named climate change denial, exploding inequality, and the movement for Fortress America as aspects of a single phenomenon:

> It is as though a significant segment of the ruling classes (known today rather too loosely as "the elites") had concluded that the earth no longer had room enough for them and for everyone else. . . . The country that had violently imposed its own quite particular form of globalization on the world, the country that had defined itself by immigration while eliminating its first inhabitants, that very country has entrusted its fate to someone who promises to isolate it inside a fortress, to stop letting in refugees, to stop going to the aid of any cause that is not on its own soil, even as it continues to intervene everywhere in the world with its customary careless blundering.

In the climate chaos we can expect over the next century, national borders as they exist today will no longer make sense (and already don't). Mexico is expected to lose half of its farms, because of climate change, in ten years, and *most* of its farms by the end of the century. Policies that embrace compassion rather than cruelty can also be climate solutions: as an alternative to treadmill economics, state or federal governments can subsidize the transition to organic or regenerative agriculture, a solution to the many problems—water pollution, pesticide pollution, biodiversity loss, economic pain, animal cruelty, and greenhouse gas emissions—created by industrial dairy farming. Many more farmers and farmworkers could be employed in dignified work, including migrants. A single policy change would do more to preserve Vermont's cultural identity as an agricultural state than would subsidies to keep the industrial dairy farm on life support.

Years ago, as a gesture of welcome and solidarity, some of us rushed out to remote dairy farms to give English lessons to migrant farmworkers. Now our children are enrolling in Spanish immersion programs, while we pore over our Spanish language grammars. We've woken up to the poverty of whiteness and linguistic hegemony. We welcome the new Americans, and our migrant farmworkers, to be a part of our community, which we envision one day will be multilingual and diverse, as it once was; we welcome them to be a part of the work of putting the *culture* back in agriculture, of restoring the farm as a place to live, of exercising our best hope.

Monarchs and Milkweed

Some years ago, when milkweed first started coming up in our garlic bed, I did not weed them out, out of concern for the decline of the monarch butterfly. The next year the milkweed was coming up all over the garden. I realized that our vegetable garden was not the best place to allow for the proliferation of milkweed. One flowerbed along the side of the house became our butterfly garden, where the milkweed plant was left alone to flourish, and over the summer, we watched the caterpillars, voracious eaters, strip them down to their bare stems. Then, we saw the chrysalises adorning the side of the house, like miniature mummies in their papery sheathes, tiny swaddled babies. One by one they vanished. And one September morning, I heard our neighbors, who live on the other side of our duplex and were having breakfast on the patio, all gasp and swoon at once. They had witnessed a monarch butterfly emerge from its cocoon and take flight.

That year, in late summer, when they arrive, and in the fall, I did see many more monarchs than I had seen in the preceding years. I had been worried, and saddened, about their disappearance. The monarch has declined by 90 percent, and though U.S. Fish and Wildlife finds their listing as endangered to be "warranted," the monarch will need to wait its turn on a long list of higher priority species. While the monarch is on the waiting list, it will receive no protection. Whenever I caught site of those amber wings, I felt joy—something alive inside of me, as if some muscle I had thought atrophied had flexed. A joy muscle. If I saw so few monarchs around our house, maybe it was because there wasn't enough milkweed. Or aster, a favorite flower of theirs. I discovered that they love the Mexican sunflower too. Maybe all I needed to do was to plant more *asclepias* and I would see more butterflies. This is something we can do to stem the flow of mass extinction—all across its range, we can allow milkweed to flourish.

A monarch born here will fly to Mexico for the winter, then in spring will fly north and lay eggs on a milkweed plant along the Gulf coast—in Texas or Florida—before dying. The butterflies born from those eggs will fly further north, laying their own eggs. By August a monarch four generations removed from the one that left Vermont will emerge from its chrysalis and will fly south, all the way to the neovolcanic mountains of central Mexico. They are the only butterflies that do this—and only the eastern monarch will make the long migration.

My own well-worn field guide to butterflies, a 1964 edition of *A Golden Nature Guide to Butterflies and Moths*, predates the discovery of the monarch's winter abode. "In the fall, flocks of Monarchs move southward to Texas, California, and Florida," the guide states. It says nothing about Mexico. (The California monarch is a subspecies that does not migrate further south than southern California.) Neither did the people of Mexico's highest mountains know what happened to the millions of monarchs when they left their *oyamel* forests in spring.

It was Toronto-based zoologist Fred Urquhart and his wife Norah who solved the mystery, who set out on the trail of the monarch's migration in 1937. It took years to develop a tag that would not wash away or impede the monarch's flight. By 1952 they put out their first request for volunteers to tag butterflies with tiny labels that said "Send to Zoology University of Toronto, Canada." Twelve responded. By 1971 there were six hundred volunteers. From these tags they learned that the monarchs don't fly at night, that almost all males die on their way north from their wintering grounds, and that a butterfly can travel eighty miles in a single day. A flight pattern emerged: northeast to southwest. They found no clusters of overwintering monarchs in Florida or Texas; instead the trail of tags led them across the border into Mexico. Finally, a letter came from Kenneth C. Brugger in Mexico City in 1973, in response to an appeal the Urquharts posted in a Mexican newspaper. "I might be of some help," he wrote.

On January 9, 1975, Brugger and his wife Cathy found the first of six winter colonies, with the help of local woodcutters who had seen the clouds of butterflies. In early 1976, the Urquharts, who were well into their sixties, climbed Butterfly Mountain, wondering if the effort might be too much. And there they were.

> I gazed in amazement at the sight. Butterflies—millions upon millions of monarch butterflies! They clung in tightly packed masses to every branch and trunk of the tall, gray-green *oyamel* trees. They swirled through the air like autumn leaves and carpeted the ground in their flaming myriads on this Mexican mountainside.
>
> Breathless from the altitude, my legs trembling from the climb, I muttered aloud, "Unbelievable! What a glorious, incredible sight!"
>
> I had waited decades for this moment. We had come at last to the long-sought overwintering place of the eastern population of the monarch butterfly.

It had been a long journey.

Has there ever been a species that has been so widely loved? One that performs no essential ecosystem service that translates into economic benefits for us. They are pollinators but busy themselves with wildflowers and weeds, not apples and almonds. They are important to biodiversity, but this cannot be what motivates all the citizen science, conservation, and environmental education groups dedicated to the monarch butterfly, all across its great range.

"Few things indeed have I known in the way of emotion or appetite, ambition or achievement, that could surprise in richness and strength," said Nabokov about his passion for butterflies. "The highest enjoyment of timelessness is when I stand among rare butterflies and their food plants. This is ecstasy, and behind the ecstasy is something else, which I cannot explain."

I have read more than a few essays by my students on their experiences as children with monarch butterfly releases. To have witnessed the metamorphosis from caterpillar to pupa to mature butterfly, and then to see a pair of wings take to the air—was something they could still speak and write about with astonishment years later.

Only the monarch migrates the way birds do, and for this the *mariposa monarcha* inspires a special kind of wonder. "It has been absolutely amazing to engage with the power of possibilities and transformation when talking about the life cycle of the monarch butterfly," said one participant in a Chicago faith-based environmental program, Migration, Monarchs, Birds and Me, that connects the migration stories of butterflies, birds, and people. "A lightbulb moment for me was when I realized that like the mighty monarch, African Americans had also migrated for survival reasons," one participant stated. "What really blew my mind was to learn that it is the fourth generation of the monarch clan that makes the trek back home to Mexico. This was precisely how it often happened in my community: my grandparents migrated, my parents came later after my grandparents were settled.... Four generations later, my family made the thirteen-hour drive to Anniston[,] Alabama, where it all began."

For the Mexican American artist Olivia Barrioneuva, creator of the installation "The Monarch," in which she released thousands of farm-raised monarchs over the Silver Lake Steps in Oakland, California, they are "a metaphor of survival, of the search for a better place to live, since migration is life." "The monarch," said one visitor to the exhibit, is "an example of how fully connected life is—all countries, all animals and plants, all organisms, all ecological systems. How quickly one little action can affect the world—one butterfly, one human, one oil well, one virus. There are so many ways to think about the metaphors of the monarch butterflies."

"Every wide-eyed child and meadow walker in the eastern United States and nearby Canada knows this colorful butterfly, by sight if not by name," writes Urquhart. Though surely this was not true of me as a child. When was the first time I ever walked a meadow? Or saw a butterfly? Neither does the monarch's migration help me to connect to my own family's immigration story, which goes back many generations on my mother's side to the Great Puritan Migration, and on my father's side only two generations, though my grandfather's story is buried in secrecy. I only recently learned that my grandfather came here with his family from Bisacquino, Sicily; that his father changed the family name from Latino to Lathem; that my grandfather had five siblings—one who was born on the ship during their crossing—I never knew about; and that he ran away from home at the age of sixteen, after he was shot in the leg, we don't know by whom or why. But like the monarch, late in my life, I am drawn to return to the place where my family story began—to Palermo, on my father's side, and to the Isle of Sky in Scotland, where my mother's grandfather was born. After four generations, is there some ancestral memory within, something that pulls and draws us home?

At the time the Urquharts found the overwintering colonies, the *oyamel* forests provided an unbroken canopy, a blanket for the fragile mariposas who require this precise microclimate, where the air is dry and not too cold and the slant of light is just right. But by the 1990s the blanket was in shreds, the habitat a series of fragmented forests that continued to shrink even after a 1986 decree that made them a bioreserve. It is no longer logging that is the threat to the monarch but disturbances at the northern end of the migration cycle. It is a widely accepted view that it is the dearth of milkweed plants—the caterpillar's only food—that is responsible for the monarch's decline, but researchers have identified the break in the reproductive cycle to appear during the autumn migration, after summer feeding. Drought in Texas and northern Mexico, a decline in the flowering plants that are the monarch's food along its journey south, neonicotinoids, and disease are all likely contributors. Tens of thousands are killed by cars and trucks. Climate instability is a threat—a single unseasonable winter storm could kill off the entire population of overwintering monarchs. As the climate warms, *oyamel* habitat is expected to shrink by up to 70 percent in the next fifteen years. (Yes. *Fifteen.*)

These are the names of butterflies. These are among the endangered, besides the monarch, who will leave behind their names like papery sheaves of chrysalises, taken by the wind and rain, when they are gone:

Karner blue butterfly, Callippe silverspot butterfly, Bartram's hairstreak, Saint Francis satyr, San Bruno elfin butterfly, Miami blue, Schaus Florida leafwing butterfly, greater fritillaries, Bay checkerspot, green hairstreaks, Euphydryas, Palos Verdes blue, gossamer-winged butterflies, blues, silvery blue, skippers, Boloria, Edith's checkerspot, Quino checkerspot, Lange's metalmark, dusky-wings, frosted elfin, Luzon peacock, large blue, Queen Alexandra's birdwing, Uncompahgre fritillary, Apodemia, Zeren fritillary, Phengaris, silver-studded blue, scrub hairstreak, Persius duskywing, Copper butterfly, Myrtle's silverspot, Plebejus, Mormon metalmar, leafwing butterflies, Heliconius, Teinopal imerialis, metalmark butterfly, red-bodied swallowtail, Bhutan glory, Wolberg, heath fritillary, branded skippers, dappled whites, Hermes copper, Atrytone, Polites, Bhutanitis butterfly.

I have still not witnessed this miracle of metamorphosis. But I have had the experience of visiting the monarch's winter sanctuary in the *oyamel* forests of central Mexico.

We climbed the trail to the top of the mountain, where we saw the first saffron wings, the little mountain kites, as they are sometimes called, swooning between the hanging boughs of the *oyamels*. As we got closer, we could see the innumerable wings stacked like pages of old books, the copper-colored covers clinging to the tall trunks of the *oyamels*. Here was knowledge gathered from the field spread over half a continent and gathered for safekeeping. They hung in clusters, like swarms of bees. Many, so many, swarms. We were just the two of us, my husband and me, and our guide, who was there to watch over us, to *guarda la silencia*. Visitors to a library of sacred books, we were not allowed inside those curtained rooms. We were not allowed more than a whisper.

We went back later in the season, when there was a sun-kissed stirring in the forests. The butterflies were waking up, dipping down to dew pools to drink, stumbling through the air, getting ready for their journey. This time we were in a great crowd, who had all climbed the long trail to see them. Has there ever been a species that is so widely loved? Here was my answer. We were all pilgrims, too, come to pay our tribute to useless beauty, and to wish these winged wonders a safe journey home.

Some, or one, four generations hence, might make it this far north, to this side of the Green Mountains, this river valley, this patio, this milkweed patch, this slant of light. We will make sure there will be milkweed here to welcome them home.

A Cautionary Tale

From our back door, on an early November afternoon, I head toward the woods at the far end of the open plain behind us, passing the garden beds of the neighboring farm, tucked for winter under cover crops, with only a straggly queue of kale still standing, and another of cabbage, left for the gleaners and the deer, who for sure will take it all in time. The brassy goldenrod of a few weeks ago has faded to a tawny din, as has the forest canopy on the hills, except for the pop of green from the hemlocks and pines. The two hills rise up from the plain like rounded loaves of leavened bread on a table, marking the end of the valley, where it so abruptly turns into a rumpled upland of steep ravines and cliffy knolls, where the sun-soaked field, pasture, and brushland becomes darkened woods.

But before the valley gives itself up for mountains, the land begins to rise and fold gently like whipped cream. I follow a mowed path along a fringe of sumac and poplar at the edge of the forest, pass beneath the power lines, and duck into the trees. A switchback trail takes me through a dense pine forest of bristly, forked trees, mixed with the slender trunks of young beeches, the trail marked with the stone steps built by youth crews in recent years, until it connects with an old logging trail on an east-west ledge. I take the trail going west, where it tilts steeply, skirting the bottom of a talus slope, a wild jumble of broken ledges and moss-covered boulders, and climb up onto a plateau. This is when I am high enough to look through the bare trees and see the sinuous rim of the low mountain chain across the valley, where the hum of the highway begins to soften. It is easy to lose the trail here, buried as it is under a carpet of papery leaves with no more stone steps to blaze the way, but my feet seem to know where to step after all these years. The path meanders and dips down into a gentle saddle, wraps around enormous glacial erratics and here and there, a giant old snag. When I reach the end of the trail, where it connects with the trails into the town forest, I loop around and head back the way I came. I could keep going forever, through the town forest, which abuts the 72,000-acre Mount Mansfield forest, together constituting one of the largest tracts of unbroken forest in the state, extending from the shores of the Winooski to the summit of Mozobediwajok (Moosehead Mountain), also named Mount Mansfield, Vermont's highest peak.

It's hard to image but some two hundred years ago this was all sheep pasture. There was hardly a tree left standing. I feel a certain awe—that is, a mixture of horror and admiration—in thinking about how, without power tools or tractors, a handful of settlers were able to fell all those giant trees, and in so little time, fueled only by a great hunger and rapacity. The stone fences they built by hand, stone upon stone, to contain their sheep and required by law, amounted to a mass of rock as large as the Great Pyramids of Egypt. All of this was done not to build lasting monuments or the foundation of an enduring culture but to extract short-term profit while the slings and arrows of the economic system were pointed in their favor, the pastures and rock walls abandoned once the tariffs on wool were lifted and prices crashed, leaving the soils degraded, and a legacy of ecological damage all that remains. The sheep craze lasted only three decades, the same length of time that Art has lived here. That is, less than a millisecond in the deep scale of time.

It is a cautionary tale.

Pastoralism *can* be sustainable (as Art has sustained our flock for more than two decades without ecological degradation), and can even have a beneficial impact on wildlife communities, as goat walker Jim Corbett has shown. But "pastoralism," he reminds us, "is responsible worldwide for the destruction of arid lands. . . . Goats [and sheep and cows] make deserts. Unless the herders practice—or imitate—nomadism by rotating pastures—and culling." In short, raising grazing animals contains within itself the potential to wreak ecological havoc and must be practiced with the greatest restraint—none of which was in evidence in the sheep craze of the early nineteenth-century Vermont.

Sheep are typically named as the driver of the deforestation of Vermont, and much of New England, although they were only a contributor. William Jarvis, the Spaniard who introduced merino sheep to the United States in 1806, igniting the sheep craze, chose Vermont as a place to distribute his prized merinos because the hills were already denuded of much of their forest. They built those stone fences because there were no more trees. In a sense, the sheep farmers were opportunists who colonized a disturbed ecology, invaders who did much damage but did not stay. The teeth and hooves of nearly two million sheep then trampled and eroded the thin and fragile soils, now exposed and no longer anchored by trees, in places down to bedrock; they encouraged further deforestation, and prevented the forests from recovering for at least a generation. But timbering for firewood, building materials, and potash began well before the arrival of the merino, and continued after the pastures were abandoned, reaching its peak in 1889, when there was too little timber left to be worth the trouble of logging.

Those forests were nothing like the forest I have just walked through. The butternuts, chestnuts, and elms are gone entirely, the giant old beech trees that could live for hundreds of years, the tall straight pines that were prized for ship masts and cathedral building. The understory with its spring ephemerals never did return. The spindly immature trees of the forest have remained so, because they are already dying. Now it is the ash trees that are in trouble. The white pines that colonized abandoned pastures are forked and multistemmed and oozing with pitch, unlike the stately giants that were the first trees eyed by the new arrivals, who labeled the white pine the "King's tree." The maple saplings, browsed by an overpopulation of deer, will never even have a chance.

As I walk these interconnected trails I think about the trail network extending throughout the Green Mountains that preceded the colonists, connecting the Indigenous Abenaki with other tribes, who also used these forests, and who took refuge here in increasing numbers as the settlers drove them from their homes following the so-called King Philip's War. These old logging roads might have been built over the ghost traces of Indigenous trails. For many generations the Abenaki, and the survivors of decimated tribal homes who found refuge here for a time, practiced agroforestry— planting many of the *pakimizial* (butternuts), *anaskamezi* (oaks), and *senomozi* (maples) from which they harvested nuts, acorns, and tapped for syrup. They harvested berries, *wigwa* (birch bark), *nbizun* (medicines), and *olakwika* (wood) from these forests for generations without degrading them. And animals—*moz* (moose), *nolka* (deer), *awasos* (bear), *naama* (turkey), *kokw* (porcupine). The colonists destroyed it all—the basis of Indigenous subsistence and culture—in three decades.

The conservation ethic that was very much in evidence throughout the Americas at the time of the European invasion, and remains so in many Indigenous cultures around the world, was not a state of innocence but was *learned*. The Clovis people, who migrated from Asia over the Bering land bridge (though new evidence suggests they were not the first humans to migrate here; others came by different routes and at different times, as Native Americans have long affirmed), and chased after the vanishing herds of mastodon, horse, rhinoceros, sabretooth cat, and other giants, were as restless as the colonists who razed the forests of New England, and had more in common with them than they did with the Amerindian or Inuit cultures who succeeded them—they produced no figurative art, no cave paintings or distinctive burial rituals, unlike the Eurasian cultures from which they sprang. The sacred ecology of Indigenous peoples, the intricate systems of reciprocities and constraints, is not a primitive form of life but a highly evolved one, representing a long and evolved adaptation to the particular restraints of geography. Refined systems of ecological knowledge, networks of sharing, and religions restraints on wastage and overhunting assured that during the eight thousand years between the megafaunal extinctions and the European invasions there would be no wave of mass extinction on their watch. The bison, the caribou, the deer, the fowl, the sea mammals, the salmon—all were relied on and none were driven toward extinction. The explanation that their technologies were too primitive for them to do much harm does not suffice, as the megafauna extinction proves, that even with the most primitive technologies it is possible for humans, at least in times of climate instability, to wipe out entire species.

A conservation ethic is the long-evolved product of many a cautionary tale. It can also be *unlearned*. The New England farmers who slashed and burned their way through the wilderness, unconstrained by limits on the availability of new land, were not as careful and methodical as the farmers of Europe who had learned to protect the fertility of their precious soils. Later the economists would enshrine the colonist's carelessness in the concept of *substitution*—which says that once a resource is used up, it can be substituted for another, forever and ever, until the earth is a wasteland. There will always be new planets to colonize.

I don't believe we will have fully settled into our ecologically adaptive phase until we are required to live with the consequences of our own lives. I question the ethics of exporting the consequences of how we live onto future generations, and onto other communities, especially Indigenous Peoples, who are losing their beloved places, so that we don't have to lose ours. Only that way will we consider that rivers and forests and soils and all the species that depend on them are *more important* than smart phones and leaf blowers and Artificial Intelligence. Only that way will we learn to practice the *code of the honorable harvest*, asking of our energy use if our purpose is *worthy* of the harvest. That is, when we ask not only about the source of the energy we use but what we are using it *for*.

The way back down the trail is the most beautiful, as I am now facing the soaring blue wall of the mountains on the other side of the valley, and in the amber light of dusk, a penumbra glows at the rim of the world, the light then passing like a hand overhead to cast the talus slope behind me in rose-colored plumage. I pick up the pace on the way down, chasing the falling light, and when I get to the bottom, where I can see the cupola of the monitor barn pointed toward the sky, our house hidden in a cluster of trees, I feel the comfort that comes with the sight of home. A pale half-moon hangs high in the sky, and behind me I hear the call of a barred owl, like a mourning call, or a persistent, repeating call of supplication, fade away as it drifts across the valley, until it is drowned out by the undying roar of the road.

Toward an
(Un)Steady State

(The Later Years)

Deer Leap

Yesterday I hiked on snowshoes up toward the power lines on the ridge. The wide old logging road climbs over the wooded hill, and before it reaches the first plateau, where the trail connects with a transmission corridor, it makes a sharp turn. Through the leafless trees I saw movement and wondered if it was skiers but then saw the deer, leaping down the wide, steep swath beneath the power lines. I watched them in their flight, one graceful leap following another in an unbroken wave along the edge of the woods. Once they passed I resumed walking and then saw another graceful dancer moving through the trees toward me, and then she froze. I did too; the antique snowshoes made of wood and sinew that I used to wear were as quiet as cat paws but not these modern varieties. I knew I could not move quietly. I could see she was a faun, who must have been left behind, distracted by some sugary residue on tree bark or beech leaf bouquet. She was half hidden behind a tree, and must have been hoping I didn't see her. I waited, not wanting to scare her into running in the wrong direction, but the longer I waited, the farther away her family would be. I decided to turn back, and hoped her family would come back for her, but when I turned around to look her way, I saw her running back through the trees again, moving further away from her family and, I feared, into a lonely winter.

Such is our disturbance, as gentle as we try to be in our step.

Those calories that family of deer burned when they ran from me were not calories they could afford, living as they are in winter on a starvation diet of tree bark and pine needles. I may feel that I am harmonizing with nature when I walk in the woods but the woods probably don't feel the same about me. The number of ramblers in the woods and on mountain trails has grown exponentially in the decades since I've lived in Vermont, with new trails appearing on mountain slopes like new lines on an aging face. Everyone else wants to take the road less traveled by, too. We live in a different world than Gary Snyder did when he wrote his poem about going off trail. Art was a serious mountain biker when the sport was new and unknown, owner of one of the first bike shops to specialize in mountain bikes. (How many can claim they have ridden both the world's smallest mountains and the tallest?) Those early riders had a code of ethics that forbade any construction or alteration of trails for their own benefit. Now Art hardly ever rides a bike in the woods. We've known to respect fragile alpine vegetation, but now I know that even on this old logging road connecting to a transmission corridor, we are a disturbance. Even this forest, worn out from historic episodes of clearcutting and high-grading, must be treated as a sensitive area if we want it to ever become one.

A little further up, past the power lines, the trail turns into the woods again and follows along a ledge with beautiful views through the trees then ends at the foot of a cliffy knoll. In the past, we sometimes liked to scramble to the top of that pile of boulders. But if there are bobcats in these woods (there are), this is surely why they are here. In a world of eight billion humans and counting, we may have to reign in our own impulse to wildness for the sake of a greater wildness around us. If we want the bobcats to come back, we will have to let them be.

Treefall

When in late winter we go for weeks
and weeks without snow, but the air is cold
enough to hurt, and a white icy crust
covers the ground forbidding an easy walk,
I am reluctant to go outside. I think too much
about winter as it used to be, the beauty
of the world after a first snowfall, before
a single footfall has blemished its perfect skin,
the next-day diamond-glitter in the sunlight,
the embroidery of turkey tracks, deer,
and the feather tip from the swoop
of a hawk's wing. And I think too much
about a future without snow. But I go out,
and having just read Ross Gay's *Book
of Delights*, I take on his challenge
to find delight. "The more you study delight
the more delight there is to study," which is
true of writing down your dreams—the more
we write them down the more there are
to write, in the way that writing poems
begets the writing of more poems. I walk
to the top of the ridge as I have done
hundreds of times, thousands maybe, before,
and as I head down, I hear a tree fall—a loud
but distant crash and then a woman
calling out to her dog, rather frantically
it seems. When I catch up with her
near the bottom of the hill, she's walking
calmly with her two leashed dogs by her side.

"Did you see the tree fall?" she asks,
and I tell her I only heard it. "I have never
seen that before," she says. These woods
are full of fallen trees—a young second-
or third-growth forest that has yet to reach
its equilibrium, with many—too many—
spindly, weedy pines and beeches
without deep roots. On a windy day
you can hear the forest creak like the
hinges on a door swaying open and closed
in an abandoned barn in your dream,
look up to see a tall tree swaying
wildly, and on one such day I did
see a tree fall: I felt the sky rip open
and a rumble underfoot and the sound
of snapping as in a great brush fire.
It lay across the path, unproud.
I had to climb over it. How rare it is
to be present when a mountain thinks
out loud. But on this day there wasn't
even a wind. It was the surprise
on her face that was the delight,
who'd seen something more unlikely
than the shooting star that passes over us
but leaves no trace, unlike the tree,
which will lie there for many years
to remind me of the treefall
that did make a sound because
the two of us met on that trail,
on a day when I almost
didn't go anywhere at all.

Spring Surprise

We were not expecting lambs. After the difficult lambing season we had in 2018—followed by a drought—we decided not to breed our sheep in 2019. The preceding spring, all the ewes had decided to take care of only one of their twins—and all of them, save one, gave birth to strong healthy pairs. Thus at one point we had five bottle lambs at the same time—a record for us, by far. And almost all the lambs were rams.

And then came the drought. In recent years we have come to expect monsoon rains and floods in summer, but that summer, the rain did not come, the grass turned brittle and brown, and the hungry flock rebelled and bolted, time and again. Art spent half his summer moving fencing and the other half chasing sheep. We had to buy hay, at great expense, in the middle of summer. That was surely the last straw—in a long series of demerits against the diminishing merits of raising sheep.

Among the checks against us as sheep farmers, our wool processer, just down the road from us, retired. And so did our shearer. In mid-May we still didn't have a shearer. The older ewes were shedding their fleeces, which hung down in long trains of ropy curls, as if they were wearing long skirts or shawls. They looked as if they had been getting undressed and got distracted.

Truth is, we were losing the war with the invasive grasses. After years of doggedly cutting the Canada thistle back with the scythe, it was at last in retreat. But what crept up on us while we were looking the other way was an invasive ornamental grass we called *jaundice grass* that started at the northern boundary of our property and then spilled over and spread like a river cresting its banks. It is now working its way toward the road. Neither mowing nor grazing is keeping this invasive in check.

We were also discouraged by the sight of all that wool piling up with nowhere to go. We no longer had the optimism we had a decade ago, in our middle years, when we were vendors at our farmer's market, selling yarn and rovings. I thought it would be so easy: I would just bring our yarn, and rovings for spinners, in their three beautiful natural colors—white, chocolate, and gray—as well as my hand-dyed yarns in a full spectrum of colors, and people would buy them. Farmer's markets, I discovered, are not where knitters and spinners go to buy their wool, and until there is a healthy local market for Vermont-raised fibers, too many wool producers will be using their beautiful fleeces for garden mulch, or will settle for selling it raw to the wool pool for less than a dollar a pound.

And then the lambs arrived. I was not all that surprised, as the ewes looked very pregnant to me (though it is hard to tell under all that wool). It is not unknown for the rubber band to fail—the method we use to neuter our ram lambs. This year, those same ewes who rejected their second-borns last year are taking devoted care of all their twins. One ewe even had triplets.

"Another lamb," Art said, coming in from feeding them one morning. I went out and found a medley of tiny black lambs scattered all over the paddock and barn, crying their heads off. I had to sort out who belonged to whom, and which one was the new one. After figuring it out, as well as identifying the mother by the tell-tail (the origin of the word?) of her afterbirth, I then wondered if there was a twin. There were still all those little crying lambs. After sorting them out, I decided one of them was the twin, and moved both newborns and their mom, Eweripedes, into the lambing pen. A little later, I returned to check on them, but there was still a little screaming lamb, unaccounted for. Could it be a triplet? I placed her in the pen with the other. Eweripedes sniffed her, approved, and then raised her chin in the air in satisfaction, as they will do. Thank you, she seemed to say. You are welcome.

I left them and hoped for the best. Would she have enough milk for the three of them? And just how, logistically, would it work? We have never before had success with triplets, who come rarely. Last year, in what was yet another misery in a long list, a triplet came as a breech birth. In our experience, a third triplet, if it lives at all, will surely be a bottle lamb.

But all three of them were standing and nursing, managing to do so all at the same time (the ewes often have extra nipples), disappearing beneath Eweripedes's long wooly skirt, with only a forest of little legs visible beneath her and their tails trembling with happiness.

I don't know why but the lambs that year seem especially sweet to me. The tiny band of triplets running behind their mom in perfect unison, or lifting up their noses with their splatterings of white, as if they'd buried their faces in whipped cream. For the lamb's life is one ongoing slumber party. They curl up together, pairs of them, black and white, nested together like inverted commas, yin and yang, or together with their cousins and the ewes, napping in the shade of the elm tree. They will run rings around a big tree, and jump, one at a time and keeping score, from the back of a resting ewe. At siesta time, they will rest, chewing their cud, with their eyelids drooping and their chins raised, acting all grown up (they do actually eat grass at so young an age).

Spring to me is the most beautiful season, and it always comes as an astonishment. The forest canopy over the hills is in full foliage—as varied in color as in autumn or more, though the colors are more subtle—from the pale lime green of the birches to the purplish emergence of the maple leaves. The black and white lambs stream in bands as they run across the vibrant grass blazed with dandelion flowers. The fruit trees—pear, apple, plum, and crabapple—are in flower. This is another benefit to postponing the lambing season until spring—the sight of the lambs on spring grass. The air smells of spring water and apple blossoms. Our world opens up from the confinement of winter in the midst of all of this emergence.

Biologists use the term *recruitment* to mean the replenishing of a new generation, from the French world *croître*, to grow, or *croissance*, growth. The lambs signify regeneration just as do the new shoots of trees. All human cultures except ours have embraced respect for natural cycles—night and day, growth and decay, winter and summer—and accepted the recurrence of life as a gift, as reliable as it is astonishing. Only with fossil fuels, which is ancient sunlight bottled and stored that we can pour from a spigot at will, have we been able to live out of synchronicity with these cycles. We have so taken this for granted that it has distorted our sense of place in the universe, and we have come to expect *renewable* energy to do the same. But cultures that relied on current sunlight falling on plants embraced a life of periodicity, of cycles of light and darkness, abundance and scarcity, and expressed this sense of structure in their cosmologies and their art.

Since that last group of lambs, we've had no more recruitment. We already miss them, but then, we don't miss having to round them up when they breach the fence. A smaller flock is less restless and not so bold. We don't miss that knock on the door: *Your sheep are in the road* (they weren't, but they were at the bottom of the driveway, licking salt). And we don't miss the pain we felt on the last day of their short but happy lives.

By now our soils have likely retired from their job of carbon sequestration, and perhaps this means we should retire as well. But we are not ready to rest altogether as shepherds, especially not Art, who still enjoys the sheep's wooly presence and the structure they give to his life. We do not have a better alternative for keeping the land open, unless we can enlist the help of a troop of mobile goats, an option we're exploring. Our vegetable garden, our fruit trees, the dandelions and clover for the bees—all need the sunlight, and so do we, it seems, as creatures who evolved on open savannas; we need the morning light pouring through our windows and the gorgeous views. If we are not producing food or yarn, it feels hard to justify this, but then we owe it to our ewes, after all they've done for us, to let them live out their lives here, the only home they've ever known.

Two ewes have passed since the last lambs were born here, with no new life to assure their forever. It feels like a diminishment, this fading away of life without recruitment. We are down to six, plus the two Soays. From time to time, Art and I have conversations about bringing in a ram to perform his *service*, for another lambing season, but then we never do. Now our conversations are about adopting a pair of ewe lambs to replace the elder ewes we lost. We've found a new wool processor who will take small batches. This could be the sweet spot between too much wool and none at all, between just enough grazers to do the job of mowing and to constitute a herd, and hearing all too often that knock on the front door.

In the Time of Corona

This is what we did during the lockdown. Ten days after we picked up a dozen chicks from the garden center and brought them home, we brought them outside for the first time. We carried them in their brooder—a shallow metal box with a sliding-top door and a floor that is removed daily for cleaning—and set them down on the grass beneath a pine tree. A mix of gray spotted barred rocks and all yellow Araucanas, they'd been spending more time looking through their windows as they grew more curious about the world outside, when they weren't in a corner huddled together as they would beneath a mother hen. We got them a feather duster and hung it upside down for them as a substitute, a little trick of simulation we learned from a friend.

We set the box on the grass beneath a pine tree and enclosed a small area around it with a piece of fencing, then opened their door and left them to find their way out on their own. One at a time they poked their heads through the door. Finally, one of the barred rock chicks jumped out of the box and onto the grass. And then jumped right back in.

Before an hour had passed they were all out on the grass, and without a ceiling overhead, they enjoyed themselves by jumping straight up into the air and spreading their newly discovered wings.

This is how we entertained ourselves three months into the lockdown. We listened to the news of hospitals overflowing, ventilator shortages, and healthcare workers without protective gear. Our own world was quietly beautiful in late spring, with so little road traffic, spring apple and lilac blossoms, and our growing seedlings in the garden. By then we'd been harvesting dandelion greens and asparagus. We had no lambs that year—but we did have our chicks.

For the next few days this was the routine—we lifted their box and carried it outside, letting the chicks enjoy the fresh air and the smell of grass, while learning their characteristic scratching and grass-eating way of foraging. That is, to express what it is to be a chicken. We knew from watching our hens raise their own chicks that in nature they would be running around all over the farm, but here they were confined to an area about the size of a queen bed. At night we brought them back inside.

After ten days we decided they were ready to move into the barn where our adult chickens nest and roost at night. We created a space for them of their own out of hay bales, thinking it would be some time before they were ready to explore the greater world of Ewetopia. But it did not take them a day.

Years before we brought home those chicks and raised them in a brooder, we had hens who raised their chicks themselves. The first time one of our hens began to set we were surprised. Over the decades, knowledge of and interest in brooding and caring for chicks has been bred out of laying hens, but some hens will still do this. Art noticed that one of our barred rocks would not get off the nest. We marked the calendar, and a few weeks later we heard the sound of chirping coming from the barn. We found the hen down on the ground coaxing her newborn chicks to jump down from the nest, which was inside a milk crate pinned to the wall at about waist height. Two of the chicks had already jumped, but the others were reluctant. The hen would fly up into the nest to be with those reluctant flyers, and then jump down to be with the others. Before nightfall she would need to have all her chicks with her so she could tuck them beneath her during the night to keep them warm and protected.

Over the next few years, that hen's daughters raised their own broods of chicks every summer, and their daughters did the same. That barred rock was the first of several generations of chick-raising hens, some of whom were better at it than others. We learned not to count our chickens . . . A hen would begin with more than a dozen eggs in her nest, but only a few, or one—and sometimes none—would emerge four weeks later.

That first brood of chicks numbered thirteen. The hen carried a half dozen on her back as she rambled all over the farm, the others following close behind. From the first hours after they hatch, chicks are truly free ranging, with rest periods when Mother Hen will gather all her children beneath her skirts, puff up her feathers, and sink down on them like a deflating balloon.

Because the hens share the nests, the brood will end up to be a mix of spotted barred rock, yellow Araucanas, and all-black chicks, from their different mothers. They are so unlike the sheep in this respect—who will adamantly refuse to care for a lamb who is not biologically their own. One year, two hens, one black and one red, chose to raise a pair of chicks together. At night the two hens nestled closely together with the two chicks beneath them. They were Mr. and Mrs. Emperor penguin, taking turns caring for their young.

There are disadvantages to letting the hens raise their own this way—for one, while the hens are setting and raising their young they are not laying eggs for us. And then there is the high attrition rate. Once, a hen sat on her nest for a month, except for the fifteen minutes when she would leave the nest to find food and water. She began with a clutch of a dozen or so eggs, but only one chick emerged a month later. A few weeks after it hatched, we noticed the hen wandering around without her chick, who we found, after a search, drowned in a bucket of water.

Of that first brood of thirteen, when they were a few weeks old, and almost ready for independence, they were attacked by a hawk. Mother Hen came to their defense, and all the chicks but one survived, though she did not.

Another disadvantage is all those roosters.

We had already been in lockdown for two months before we brought those chicks home. This is what I wrote in April 2020 when I listened to the daily news with held breath, expecting chaos. Instead, a chill settled over the world, the greatest silence since the invention of the internal combustion engine introduced the noise that has been with us ever since.

It has been more than a month and I have been to town only once, to mail some packages. Living in the country, the stay-at-home order does not feel so confining, only strange. To live inside a circle of life, watching the seasons turn, apart from history, which comes to us on the radio not like a distant foghorn but like a blaring siren inside our living room. I cannot shut it off. I talk with family members almost every day who are sheltering inside their Brooklyn apartment, and feel as if my hands were bound. I would endure this time of seclusion with more grace if I did not feel so useless, so removed from the overflowing hospitals and the city of my childhood where so many are sick and dying.

We are evaporating sap on our woodstove as we do every year, planting seeds indoors for this year's vegetable and flower gardens. The end of winter always comes slowly, but this year time is distended, fraught by strangeness and necessary hope. We had a few days when the sun was out, the crocuses up and early daffodils; we turned over the compost piles and cleaned up winter's debris. And then a cold rain came. More darkness, more winter.

Usually at this time of year we are lambing on our farm. But we chose not to do this anymore, at least for a while. I did not know, of course, when we made that decision that come spring we would be forced to stay home; that this would be a time when new life would have brought much-needed delight. More life. I look out at the paddock where I should see tiny lambs standing beneath their wooly mothers, their little tails trembling, while the ewes raise their chins into the air, as if whistling a familiar tune. I miss them.

Yesterday we watched a hawk in front of our house devour a pigeon, the snow beneath them splattered with blood and feathers. It feels like the humans are all gone and the wild creatures have their world again. The hawk had been hanging around the paddock for days watching the pigeons who live under the eaves and are very numerous, gathering in large groups on the slate roof if there is no snow, but on this day the roof was snow-covered. The hawk jabbed at the slashed-open body with its beak, tore at feathers, looked up to search around him, then went back to jab and tear again. He kept lifting his head, searching around to see if he was being watched, but he never saw us there, looking through our picture window, our cat sleeping beside us. Then he swung around and must have seen us, or the shadow of us, because he lifted the mangled bird in his beak and carried it away towards the woodshed.

The good news is that CSA subscriptions and farmstand sales have doubled in recent weeks. Local food, honestly grown, is something we can trust, and if we want our farms to be there when we need them, we better support them now. Meanwhile, the mega-dairies are dumping milk as prices are in freefall. I am reminded of those images from the Great Depression of mountains of surplus grain, farmers burning corn and, yes, dumping milk by the roadsides. These became our symbols of the cruelty and insanity of unbridled capitalism, of allowing the Invisible Hand to be at the helm. The Vermont legislature is considering emergency rules for putting a floor on milk prices—so that prices don't go below a farmer's cost of production—and for distributing milk surpluses to kids, food pantries, and shelters. These were the kinds of policies put in place by the New Deal, and over the decades since have been dismantled chip by chip. And here we are again.

It is hard to see how we will ever go back to the way things were. Putting a floor on milk prices so a farmer doesn't lose money by working seven days a week to feed us, and buying agricultural surpluses to feed the hungry—is this something that will ever make sense not to do? Will we go back to mass evictions of poor Black single mothers and their children? Throw the homeless back on the streets? Can our system of medical apartheid really continue after this? This pandemic has made visible the faults in the system, the gross inequalities, so that they can no longer be denied.

The other day my husband was looking through some boxes for some old photographs and found a newspaper clipping from 1979: "Use of fossil fuels called threat to world climate." The story appeared on page D1 of the New York Times. Even a magazine printed a few weeks ago *arrives in our mailbox as if it were sent to us from a distant past, as did this newspaper clip. The world before corona. I worry that the climate threat is once again pushed back to page D1, although we know now that the effects of fossil fuel burning will not wait two hundred years, as the professor thought in 1979. That they are already here, and that climate change will bring more pandemics, more economic disaster, and will kill many people. We will have to reckon with the knowledge that, once again, it is only in times of economic contraction that emissions decline.*

I worry that the climate threat will always be pushed back to page D1 because there is always an immediate crisis, because people need to eat, to work, to live, while the earth is more and more diminished and ecologically disturbed, meaning that there will be more crises, more unmet needs. A forest, writes David Quammen, "with its vast diversity of visible creatures and microbes, is like a beautiful old barn: knock it over with a bulldozer and viruses will rise in the air like dust." We have been warned.

Friends and family members tell me they are rereading those plague-themed books, like Camus' La Peste, and Love in the Time of Cholera. I am rereading Angels in America, and grieving for the many who died of AIDS while the world was indifferent. We did not, as a collective, stand in our doorways to bang pots and pans at seven o'clock. The stars did not go out, we did not pack up the moon and dismantle the sun. I want to believe that the near universal solidarity we are living right now will not fade away, that it will change us, deepen our powers of empathy; that the circumference of our sphere of care will expand in time and space, to include the past, the far away, the seventh generation, the neighbor we never knew. The animals, too, who are our permanence, our forever.

For a moment there the world's response to the pandemic seemed to show that we *could* just stop Business as Usual to address a global emergency and that the public would cooperate. But as a precedent the COVID pandemic turned out to be a terrible one, proving that people will find imaginary, tyrannical plots under *temporary* health measures, and once again science will be rejected for the wildest of conspiracy theories. The homeless would indeed be back on the streets and in greater numbers than ever before. And the ones who would suffer the most long-lasting damage from the lockdowns would be the children.

That chaos did come. Our divisions and inequities only widened, until we no longer even shared a common reality. The divide was more than political or ideological—it was existential. "If Biden wins, the world is over, basically," said a Trump supporter quoted in *Time* magazine during the election season of 2020. "I will probably take my children and sit in the garage and turn my car on and it would be over," she said. I often heard dire language coming from those who feared another four years of Trump, although I never heard anyone say that they would gas their own children if he won.

The downward curve on our emissions would soon be reversed, and just when it seemed we were making forward progress on our transportation issues in town, the wind was knocked out of our momentum. Our limited bus system between Vermont's largest cities would never recover its pre-pandemic levels of ridership, and the neighbor-to-neighbor ride sharing program we were about to launch—no app required—never did get off the ground. (For years our town has studied the problem of how to get people from our village center to the bus stop and the park-and-ride, when every minute of every day a stream of cars with empty seats flows on by.) Our renewable energy goal posts just keep moving further and further into the distance with the arrival of new digital technologies—bitcoin alone has consumed more energy than medium-sized countries, erasing all the gains the world has made with building solar energy. Now, with the rollout of artificial intelligence, the technology companies are planning a massive expansion of data centers and are firing up coal plants and planning new nuclear energy generation to fuel them. In time, American voters would again elect a man with so much hate and fury he would pledge to uncork the stops on carbon pollution and bring the world down with him in flames.

One by one the chambers allowing air flow to our organs of hope are closing.

I chose this life, I wrote, *because I have never had much confidence in a precarious, bloated economy of global supply chains, credit default swaps, technological dependency, two-thousand-cow dairies and over-specialization. Instead, it is the rhythms of a rural life—collecting sap in buckets, cutting wood for the stove, caring for animals—that are more reliable. When everything else is failing, these do not fail us, cannot fail us. Animals, who have always been with us, are our link to the past before the past. By caring for them, for the soil and the forests, the carbon and nitrogen cycles and the rain, we are keeping forever alive and this is surely not useless.*

We did not entirely go back to the way things were before the chaos of the pandemic times, which did create an opening for a national racial reckoning in the wake of the brutal murder of George Floyd on May 25, 2020. It was as if white America (but only, as it would become clear, a part of it) was shaken from its long spell of illusory color blindness. Though we are also a farming community, this town is functionally a bedroom community, grown into a suburb of Burlington, with all the impediments to greater inclusivity and affordability—no public transportation, almost no mixed-used zoning, scarce affordable housing, no mosque or synagogue. Our migrant worker community lives in hiding—when not even during the time of the Underground Railroad did refugees from southern plantations need to live in hiding in Vermont (contrary to the myth) but lived and worked in the open. In 2020 and 2021 some forty participants attended the weekly meetings of our local grassroots racial equity group, and although that number has shrunk by two-thirds, a core group remains committed to doing the necessary work over the long haul, and because we are one of the whitest states in the union, it will be a long haul.

We brought home a dozen chicks that pandemic spring because we thought not all of them would survive, and chances were good that some of them would turn out to be roosters (chicks are notoriously hard to sex.) Two years later we still had all twelve, who were all hens, as well as our older black hen, Lady Macbeth. This group stayed close to home, never went near the road, and did not cross over the third rail into the neighboring farm. We guessed that in the past it had been the influence of our big bad roosters who encouraged the chickens to help themselves to the tomatoes and cucumbers next door, and we kept our fingers crossed that this would not be a problem again.

They gave us a dozen eggs a day, more than enough for us and our neighbors and friends. Hens who are truly free range, who live on grass and bugs and are not grain-fed, have far better eggs than any "organic" or "cage free" eggs you will buy at the store, with yolks the colors of the most brilliant setting sun (a sure sign they are eating grass). When eggs come from truly free-ranging, pasture-raised hens, they are the most sustainable form of protein that I know. It is the most gifting source of nourishment for a small investment of labor, and in our household it is our staple protein, rich in vitamins and omega fatty acids as well as protein. Deviled eggs with an array of salads and pickles on warm summer evenings. Omelets—yes—with leeks. Egg salad with fresh dill. Eggs cooked in a nest of tomatoes and chiles. Eggs for pancakes, scones, muffins. No suffering, no wrangling, no carbon pollution involved. I do believe that backyard chickens and kitchen gardens—Huertas gardens—will be a key to affordability, and to the successful resettlement of New Americans, many of whom come from farming cultures. In the event of a bird flu outbreak spreading like wildfire through industrial poultry feedlots, backyard chickens, who are healthier and more resilient, can be a source of food security.

Now we are into our fourth summer with this new generation of hens. They are keeping close to home and have shown no interest in the rows and rows of beautiful vegetables that the youth crews are so carefully tending behind us. Our chickens do not need us to feed them; offerings of grain and kitchen scraps at random times during the day proved a better method of keeping them closer to home and out of trouble than attempting to fence them in. We have lost only one of the original twelve we brought home in a box during that strange time. We are blessed that our chickens are not preyed on by foxes or other predators. I know that where conflicts arise between livestock and wild animals it will be the wild creatures who will suffer in the end. To protect them, farmers will have to shelter their livestock and limit their freedom to roam. But we are somewhere in that sweet spot between not enough open land for chickens to range freely and too much to keep them safe. Maybe that is the holy grail, the fine and invisible line between the domestic and the wild, between too much control and too little.

Listening to Loons

St. Regis Canoe Area, Adirondacks. When we arrived there was a strong wind coming off the lake. It was not what we expected. The surface was ruffled but we were used to canoeing across open water in much rougher winds than this. We were hungry and decided to eat before we loaded the canoe and headed out. We peeled and ate two hard-boiled eggs and some crackers with sheep's milk cheese, sitting on a log by the put-in with our backs against the wind. Then we loaded our big pack with our clothes, cooking pots, sleeping gear, food bag, and water bottles.

On the map, the portage we were looking for was more or less directly across, but we could not see anything that looked like an opening in the trees or the old railroad bed. Across the lake a small mountain rose up in a perfect cone, and above us a few bright clouds floated in a crayon-blue sky. There were no boats on the lake. We passed a single loon, whose periscope searched the area, spotted us, and then dove under. The light shone off the wind-stirred surface in silver threads. I dipped my paddle into the deep water, leaned in, and then leaned back, repeating the motion in a steady rhythm. After some time, I switched to paddle on the other side and could feel my husband behind me follow my lead. We paddled to the western shore, having spotted what seemed like an opening, but it was only weedy swamp and no sign of a trail. We saw only trees, white cedar and spruce, and a dense wall of evergreen circling the lake, with no gaps, and we were beginning to wonder if we were going to find the portage. We came upon a group of loons. Then we circled around into a bay and found the old railroad bed, where the trail is.

The portage was very steep on both sides of the railroad bed. It was covered in loose gravel that made it difficult to get a firm foothold. There was a culvert with a stream of water rushing through it. We pondered it for a while, then decided to try to pass our heavy canoe through the culvert instead of carrying it over the trail. Art carried the pack over the trail and I waited at the other end for the canoe to come through, when I would have to guide it over a foot-high drop. I stood in my flip-flops on the slippery rocks, before a lilliputian waterfall dropping from the culvert. Art guided the canoe, bent over at the waist, because he could not stand up inside the culvert. "I have it," I said, as he stood at the end, afraid to let go. "You should turn around and walk back."

I stood in the shallow water among the pickerel weeds and water lilies, and then guided the canoe to the sandy shoreline. We loaded the big pack and climbed into the canoe and began paddling across the long narrow pond, keeping an eye out for a place to camp. We saw a canoe pulled up on the shore and through the trees we could see a blue tent. There was no one else on the pond, which was hemmed in by a wall of cedar and fir trees, and along the banks a dense mat of sheep's laurel bushes that I first mistook, from a distance, for blueberries. The sheep's laurel flowers were in bloom, pink and cupped. There were no boats on the water, and we saw no opening in the forest. It was a while before we found a campsite. We pulled up and walked a short path through the laurel into a roomy space with a floor carpeted in pine needles, raised into pillows and mounds, but there was one flat area where we could pitch our tent. Beside a fire pit someone had piled some branches and twigs and in the fire pit was a half-burned log. We put up the tent and then we got back in the canoe to explore the pond.

When we got back we opened a bottle of wine and poured the wine into our two enameled cups and drank it seated on a large log facing the water. We sliced some of the sheep's milk cheese with a Swiss army knife onto crackers and ate that with the wine.

We sat there with the light of late afternoon shining on the water. We heard a fish jump, and then another. We watched a kingfisher vaulting back and forth across the pond, from shore to shore. We heard a quarreling among the blue jays.

A red squirrel scrambled up the trunk of a tree beside us, then stopped and rattled his noisemaker, with one eye cocked toward us. "O yeah?" Art said. He often talks to animals.

When it was time to make dinner, we started up our little camping stove, drew some water from the pond, and placed a pot of water onto the stove. I cut up some kale, green onions, and sorrel I had brought from our garden. When the water was boiling, we added noodles, spices, tofu, and then the greens and boiled it all together until the noodles were cooked.

We started a fire in the fire pit after we finished our dinner. Art broke up the branches and twigs and gathered some pine needles, and when he had a good blaze going, he put on some bigger pieces, and then he rolled the half-burned log onto the flames.

We watched the fire, throwing more wood on as it died down. The loon began to sing—that long plaintive song that seems to drift so slowly through the world's loneliness like something falling through water. A bullfrog bellowed through the night, brassy and loud.

I slept soundly on a comfortable air mattress in my down bag and woke to the explosion of birdsong at dawn. I went back to sleep and when I woke again it had quieted down. A loon passed by tooting a new song, a different one than the melancholy song of the night.

We cooked oats and made coffee and ate the cereal with fresh strawberries we brought from home. After breakfast we set out in the canoe. The pond ended in a shallow bottleneck that joins another pond. The surface of the water was a mosaic of heart-shaped leaves, the long grasses that lay flat on the water as if the wind had blown them over, the straight spear-shaped leaves of the pickerel weeds spearing the sky, crowned with purple flowers, the tight round bulbs of the pond lilies. We paddled through the weeds and flowers that brushed the sides of our canoe as we passed.

We saw a loon ahead and so we moved slowly and gently. It dove and disappeared, we paddled forward, gently, and then it popped up just ahead of us. We slowed again. It dove under, we paddled forward, and it popped up again. It turned its body so that its red eye was locked on us, loons being cubist creatures of two dimensions. This went on for some time. It was a kind of duet between us. We had come so close to the loon that we could see the intricate pattern of its coat, a checkered blue and white, the iridescence of its feathers, and its ruby eyes. Behind it the yellow heads of pond lilies nodded, and the pads of lilies spread out over the still surface. Then there was a rustling in the bushes, we turned our heads to look, and so did the loon. It dove again. And then I saw it swimming underwater, alongside our canoe, its body pointed like a spear.

We scraped bottom and so I got out and pulled the canoe over a shallow bar, where my paddle caught in the tangle of weeds, and the willows brushed against us as we pushed our way through. Then we were gliding on open water.

We looked for the next portage, and had to circle around, hugging the shore before we found it. This one was not steep but it was long. We emptied the canoe, and then carried it, one of us at each end, stopping to rest our aching hands from time to time. There were birches in the woods there, big old birches with curling yellow bark and gnarly limbs. We came out of the woods onto the lake. The sun had come out and I could feel it hot on my bare skin. It was a large lake shaped like a tree, with one central trunk and several arms that we would paddle along, moving up and down one and then another. A couple in two separate guide boats passed and we nodded in greeting but didn't speak. We paddled to the end of a bay until we had a good view of the mountain we'd seen at the put-in the day before. It was closer now. We hugged the shore as we looked for a place to stop to have our lunch, and then we came to an empty campsite. We pulled in and sat down on the pine needle carpet and ate boiled eggs, bread with cheese, and some pistachios. After we ate, we paddled to the end of the lake, passing another pair of canoeists, who nodded to us and we waved in return. I was growing tired as we headed back, carrying the canoe over the long portage, and at the end of it I took off my clothes and jumped into the cool water to refresh my bones and wash away the pain and tiredness. There was no one anywhere and I didn't see any need to change into a suit, so I didn't.

That night we drank another bottle of wine, and ate the last of the sheep's cheese, and then we cooked rice and lentils, seasoned with a packet of onion soup mix. We built a fire and watched and listened to it, and we listened to the loons and the bullfrog, and in the morning the explosion of birdsong and then the quiet that followed.

It rained when we paddled out; the lake was gray with mist, the forest whispered, and the rain made pinpricks on the water's surface.

When we got back to the put-in where we'd started out two days before, the same loon appeared who we'd seen when we headed out. It dove under as we neared. The water closed around it like silence around a word after it's been spoken.

The Great Flood, and Other Catastrophes, 2023

On the morning after torrential rains during the night, I looked
east toward Waterbury from our front porch at what looked
like a lake where there should have been a hayfield. I hopped on my
bike and rode as far as I could, along the edge of the flooded field,
past a small waterfall where the water dropped over a ledge I'd never
even noticed was there, and where I saw someone practicing his
river-running skills in a kayak. The water had crept across the road
in the center of Jonesville, just beyond the ReFind Boutique and
the flea market. There the road had been closed, and if you hadn't
known better, you would think it simply disappeared.

When I got home Art had just come back from riding in the
other direction, toward town. "They've lost another tractor," he said,
referring to the last time Maple Wind Farm's pastures were flooded,
when a freak Halloween-night storm drowned their tractor as well
as hundreds of chickens and turkeys. We went into town together,
past Maple Wind's flooded fields (what we used to refer to as the
Andrews' fields), where I saw the top of the submerged tractor
trailer, left there as if it were suddenly abandoned, which it was. I
saw the hoop houses where the chickens live, and hoped that they
had been evacuated before the waters rose. I would later learn that
the farmers had moved their poultry from the lower pastures in

anticipation of flooding but had not expected the waters to reach so high. In town, the river had crested the banks and flooded the town green, the playground and ball field, and the garden in front of Stone's Throw Pizza, where the tops of the tomatoes clung to their trellises as if they were trying to climb to safety. The water had crept around the back of the Town Center, behind where we stood, among the other townspeople who had come to marvel at the sight before us. The water was moving fast, turning over at the edges in little waves, forming ripples of light across the surface of this shallow lake. The water poured from the road and onto the bridge, which had become a bridge to nowhere. One poor deer would be caught on camera being swept across the road by the fast-moving water until she was able to climb onto terra firma where the waters had begun their retreat. She rose, looked around, and then ran away from the rushing waters that had trapped her. Esplanade Street, with its row of houses along the park, was underwater.

And just as quickly the water receded.

And just as quickly the town sprang into action. We had been through this before, eleven years before, to be exact, when Tropical Storm Irene had flooded the village in what was the worst flood in Vermont history since 1927. In the summer of 2011 we'd already had two flooding events before Irene hit, leaving the ground so saturated that when the rains came the ground could not take in any more water. We were devastated as if by a hurricane but this was merely a *tropical storm*. There was not even much wind. At the time, I was unable to be a part of the cleanup because I was holed up with a swollen knee and was on crutches. (It turned out that I had Lyme disease, also a consequence of climate change, but it would be two months of hobbling around on one leg before I would be diagnosed.) All that I was able to witness of that disaster were the flooded fields I could see from a car window—the inundated corn fields, the marshmallow hay bales carried two miles downstream to be deposited near the interstate exit.

Farming has always been risky, but now it is much more so, and I wondered who would be willing to take on such a gamble in these times. After Irene, some Vermont farmers quit, having suffered three flood-strikes in a row; others, once they recovered their sea legs, came back the following spring to plant sunflowers and carrots and lettuces, and to put their new lambs and calves out to pasture, beginning again—as refugees will do, continuing the cycle of life as we have always done.

This time I was ready to roll up my sleeves.

This was the second thousand-year flood in a decade. The Winooski rose even higher this time than it did in 2011. We learned a lot from the previous flood, and this time the town was prepared. Mitigation measures put in place after Irene had worked, saving some towns across Vermont from being inundated. Some had raised their houses, with help from FEMA. Residents of homes in floodplains knew by now not to keep anything valuable in their basements. One resident organized the volunteer response team with a spreadsheet; she had a pop-up tent set up in front of the Town Center, with cases of bottled water, snacks, cleaning supplies, and rubber gloves. Volunteers could just show up and she would tell them where to go. The congregational church served lunch and dinner, where volunteers showed up to cook and clean and others dropped off casseroles, lasagnas, and baked goods.

It was hot. It was humid. The steam seemed to rise up visibly from the ground. I headed over to Esplanade Street, where residents had emptied the contents of their basements and piled them up on front lawns. Crews of volunteers were hauling it all to dumpsters. Old skis, hockey sticks, leftover flooring, boxes of jars, half-empty paint cans, electronics. In front of one home, with the biggest piles, the one that seemed the most distressed, I saw a teenage girl carefully laying out wet photographs on the grass to dry out. But mostly we were tossing stuff that didn't matter, the kinds of things that fill America's basements. *We are coming to the end of the Age of Basements,* I thought. We will have to learn to live more lightly, in readiness to pick up and go as the waters rise and the fires rush to claim our homes. We will have rehearsed the question: *What will you take?* We will be experts at evacuation, as we already were, practiced at picking up and moving on, carrying only what we need.

We looked up at the sky and braced for more rain.

The year 2023 was the hottest on record in Vermont and in the world. All that increased energy in the system means a lot of wild weather, and new records were broken. We got used to hearing the word "unprecedented." After a very warm winter, our first of three weather-related disasters to qualify for federal assistance was a hard freeze over the night of May 17–18, just that time in the season when fruit trees and grape vines are in blossom, when the tender flowers are vulnerable to frost. "In my 25 years of working with fruit crops in Vermont," said Terence Bradshaw, associate professor at the UVM Extension Fruit Program, "I have never seen frost or freeze damage this extensive. The widespread nature of this event is unprecedented, and has affected orchards and vineyards across the state."

Alerted by the forecast, we covered our onions, beans, and lettuces, which were unharmed. But fruit trees are not so easy to cover. Entire orchards, impossible. This would be a year without an apple or pear harvest for us; far more devastating was the loss for vineyards and apple growers all over Vermont, and for the guest workers who rely on this seasonal work for their livelihoods.

Then came the season of smoke from forest fires raging in Quebec. The heavy pall of smoke that blanketed huge swaths of the Northeast bypassed most of Vermont, New Hampshire, and Maine, while it smothered New York and much of the mid-Atlantic and Midwest. Still, a dull haze did hang over us for weeks, when we made it a daily practice to consult the EPA smoke alerts; on some days we were advised not to exercise outdoors. It was strange thing to have sunny days with no blue skies.

The massive pall of smoke that smothered almost half a continent was a measure of a fire season that exceeded all fire records by a "stupendous" margin. Over a thousand megafires burned across Canada that summer, with more than six hundred fires burning out of control for weeks. The largest of the fires in Quebec began on June 1 in Eeyou Istchee, the land of the Cree, near the La Grande

River and the site of Hydro Quebec's massive James Bay I hydro project, which had flooded 11,550 square kilometers of forest. Now the Cree who mourned that loss had lost an even greater area of forest to fire. Scorching 12,250 square kilometers, that one fire burned two and one-half times larger than the biggest wildfire in California history. Eight out of nine Cree communities had to be fully or partially evacuated as their skies turned red and orange and filled with smoke, as the flames nearly licked at the edges of their villages.

And across Canada some twenty-five thousand Indigenous people were evacuated during the course of Canada's longest and worst fire season. It was worse by several orders of magnitude. Look at a chart comparing 2023 with previous years and the bar that shoots up in 2023 looks like a tall tree compared to the little stumps that precede it. By October, 6,551 fires had burned 184,961 square kilometers, an area twice the size of Portugal. Many of those fires continued to burn underground as slow burning "zombie" fires that threatened to jump-start next year's fire season. "Never before has a snowstorm smelled to me like smoke," said one fire fighter in British Columbia *in February*. "People talk about the fire season ending. Our fires dug underground, and have been pretty much burning all winter."

The Peoples of the boreal forest, like the Cree, have always lived with fire—but not like this. Climate scientists think that the forest fires in Canada in 2023 indicate a shift, a change in status. That is to say—a tipping point. Monstrous fires on such a scale, and over such a vast region, have changed the environment and that means not only the boreal forest but the carbon balance that gave us some ten thousand years of climate stability. Carbon emissions from Canada's 2023 fires were comparable to those of a large industrial nation, and were surpassed only by the three largest emitters (China, India, and the United States). Plumes of greenhouse gases from fires in these amounts, if they continue, will by far outpace our ability to reduce emissions, even if we meet our most ambitious targets. Forests that have evolved in adaptation to ice will now have to adapt to fire.

This too was not something I ever imagined I would see in my own lifetime. Canada was supposed to be one of the safer places to be. Like Vermont. It was supposed to be a place of refuge.

Here in Vermont, those of us who think of these things are thinking ahead to an influx of climate refugees. Vermont will be a safer place to be only because we do not already live close to the climate edge. We have a wider climate margin to adapt to a warming Earth. But here in the Northeast the climate is moving faster than it is anywhere else in the contiguous United States. We are on track to reach the limit of 2 degrees C (3.6°F) of warming two decades before the global average. That is by 2035—only a decade away.

We had still more records to break before the year 2023 came to an end.

It was still brutally hot and muggy the day after the floodwaters receded. I joined the cleanup at the ReFind Boutique in Jonesville, an antique and upcycle shop owned by a Ukrainian Canadian who had dedicated a portion of her store for donations to Ukraine. She had raised tens of thousands of dollars. On this day she still wore her big smile with adjoining dimples, but there was also dread in her eyes and tension, and her face was flushed from heat. I joined another volunteer who was cleaning the crud off of a set of chairs with a rag, and started to work on a table. "That was underwater," JoAnn said. "It gets tossed." She looked over the pile of furniture beside me. "That is wood underneath," she said, pointing to a wardrobe. "It can't be saved." "Mold on wood is a hopeless battle. Toss it." "All the electronics get trashed."

What we call Jonesville is the ghost of a town center, about a mile up the road from our place, that was once bustling (it was never incorporated as a town and is officially a part of Richmond). Today, Jonesville is just a cluster of half a dozen buildings beside the railroad tracks, right before the bridge over the Winooski. Photographs of its glory days show a busy railroad depot, a four-story hotel, a large sawmill that before it closed in the early seventies turned out three and a half million board feet of lumber a year, a wood shop that manufactured whole houses, a grocery store, a post office, and a school. Today the old schoolhouse still stands, though it is no longer a school, and the warehouse that was the former Plant & Griffith Lumber Company hosts a flea market, a small cabinetry business, and, until recently, a gun shop.

Jonesville has been through its whirlwinds of change. The 1927 flood took out the old covered bridge and swept away two homes but spared the schoolhouse that stood next to them. The lumber shop was rebuilt after being partly wrecked in the flood, then was rebuilt again after it was partially destroyed by fire in the 1950s. The sawmill closed in 1972, when the company started importing its timber from Canada and the Pacific Northwest and continued to turn out roof trusses and entire homes equipped with prebuilt cabinets until 1976.

Jonesville is haunted by ghosts that have a story to tell, one that played out all over New England and across the continent as manifest destiny moved like a shadow from east to west, taking the forests with it. The 1927 flood remains the most destructive in Vermont's history because it was not so much a natural disaster as a human-made one. A photograph of our farm from a century ago shows the hills around us altogether denuded of forests. All over the region, as New Englanders engaged in a frenzy of timbering, without the forests to absorb water and anchor soils, floods and mudslides would scour the topsoil from valleys and hills, taking out bridges, mills, waterwheels, dams, homes, barns, and anything in the way of the raging waters.

It is seductive to imagine colonial Vermont towns as self-provisioning communities of homesteaders. The settlers did build their own homes and outbuildings, made their own clothes, and grew their own food, and for this I admire them. But a rapacious extractive economy was at work in the years that settlers razed the forests, dammed the rivers, and exported the region's natural wealth in the form of potash, lumber, grains, livestock, and wool, and, when the wool market bottomed out, dairy. This is one way to understand what Marx meant when he famously said that capitalism will in the end destroy itself, not because the workers will revolt and build a socialist utopia but because the earth is not a perpetual fountain of resources for the taking.

Milldams and waterwheels for sawmills and gristmills were erected on any and every suitable stream—in our town, on Snipe Ireland Brook, Jonnie Brook, and the Huntington, especially the Huntington. The Huntington Gorge is now a beautiful swimming hole, and the water runs clean of sawmill waste. It is the rivers elsewhere that produce our electricity, and absorb the waste from the mills and factories that make up the stuff of our lives. Mostly, all that we export now are consequences.

But the consequences have come full circle and are here, too.

We have our forests to thank that the damage was not much, much worse. A single one-hundred-foot tree can take up eleven thousand gallons of water in a season—with one to two hundred trees in an acre of forest (that's over a million gallons), they're holding back the deluge.

Next to the ReFind Boutique is the flea market housed in the old Plant & Griffith Lumber Company building. Out front there was already a small mountain of stuff that had been ruined, which all had to be lifted up by the sodden armload and tossed into a dumpster. Old paintings and prints. Boxes of record albums. Toys. A carboard box full of "SALE" stickers. Electronics. We stopped when the dumpster was full though we still hadn't dug through to the bottom of the hill.

It had been a while since I had been inside, as the flea market hasn't been open for retail sales since the pandemic. It is a place where Art and I have gone for tools, and I have often enjoyed just perusing the collection of antique farm implements, and all those hammers and drills with their well-worn handles. It is where I found my scythe. But by now the interior was so crammed with stuff, stacked up in piles taller than I was, that you could scarcely squeeze through the aisles. I thought of what JoAnn had said: "Mold on wood is a hopeless battle." I thought of how easily the owners of the homes on Esplanade were willing to toss things. Here was a warehouse full of stuff that had already been thrown away at least once. Over the next two weeks when I'd pass by I would see them still at work, carrying things from one bay to another, tossing stuff into a dumpster.

Only one week after the July 11 flood, we had torrential rains again. I listened as it pounded the roof and gushed from the gutters, as it hissed when it hit the puddles that had so quicky formed. Art went downstairs to check the basement and found water gushing through the cat door like a firehose. The floor was inundated up to his ankles. We spent the evening vacuuming it up and carrying bucket after bucket of water outside to dump onto the already saturated ground. We knew better than to store anything on the basement floor, so there was no damage. But the damage to Richmond roads and driveways from this event turned out to be even worse than the flood the week earlier, amounting to millions of dollars in road repairs. It was a one-two punch, the second storm affecting homes left unscathed by the earlier storm, including homes that were nowhere near the floodplain.

Over the next weeks, and months, the rain never stopped. Mostly it was just a drizzling rain, but it was a rare day when we saw blue skies. Our gardens benefited from the rain to a point and then they suffered. My cucumber vines turned yellow and died. Onions should be harvested in dry weather but that weather never came. The same with the garlic. A plague of mosquitos like I have never experienced here kept us from working in the garden, and so we neglected it. Snails were everywhere.

In autumn, we waited for the brilliant fall foliage that never came. The rains continued. Only for a stretch of days in September did we get a break, to soak in those glorious glowing days of early fall. But by October the rain cloud had returned to stay. The leaves on our sugar maples went brown and dropped, never turning their usual brilliant gold and amber. We expect our hybrid maples to explode in a burst of bright red foliage, but this comes later in the season, so I was patient, and hopeful. I watched as a nimbus of rust rose through the canopies like a dud firework. Throughout the landscape—with the exception of a pocket of color in Hinesburg—the hills were covered in a mud-brown canopy, like a blight had passed over the world, which it had.

A fungus was responsible, called anthracnose, which thrives on wet leaves. Another factor in the lack of color was too little photosynthesis from so little sunshine, which meant the trees produced too little sugar. I had thought about and studied climate change for decades and imagined I knew what to expect. The maples would in time creep northward but they would not pull up their roots and walk away in my lifetime. I worried about beetle infestations, which were responsible for Dutch elm disease and the widespread destruction of pine forests, as were tree caterpillars and spruce bud worms. The ash borer was here to stay. But this—a fall foliage that would not come—was never something I imagined.

No one really talked about it. Everyone seemed to be thinking about it, but we were mum. It was all simply too much—after what we had been through. The news media did report on the "disappointing" fall foliage but emphasized that the leaf-peeping tourists should not cancel their plans. *You can still find color if you know where to go*, they said. Fall color is not something you should have to hunt for. That was the beauty of it: it wrapped its arms around you like a warm day. It fell upon you like a blizzard. It held you inside of its whirlwind until you were unspooled of yourself. It crushed you with its beauty like a Mozart or Mahler symphony and you left your old self behind.

"We have good foliage years and others are not so good," is what I heard from friends who are well aware of climate change. I have been in Vermont for thirty years and have never seen an autumn as dull as this one. In some years we have more sunny days, when the light brightens the colors, others when a single blustery October day will knock down all the leaves before you are ready for it to end. But this was something else.

The Northeast Climate Assessment reports that researchers expect an *increase* in fall color as the region warms. The anthracnose fungus was probably not figured into those climate models. Still, it gives me reason to hope that this is not what has become of autumn in Vermont. The fungus is not here to stay. There will be rainy years when the blight will pass over the land; there will be others when the colors will be brighter than anything we've ever seen.

Moving Rivers, Phantom Houses

After the big storm, Art and I rode over to visit our favorite swimming hole to survey the damage. From the road, the landowners have generously maintained a path for public access to the river, passing along the edge of a lush green pasture that was now unrecognizable. The flood waters had buried the pasture and left an ugly expanse of sand and gravel. There was not a living blade of grass where there had so recently been a splendor. A riparian buffer of shrubs and small trees had been ripped away to expose a changed river, with new gravel bars and islands and a greatly expanded beach. Across the river we watched as a small group had formed around what appeared to be an old merry-go-round that the river had carried from who knows where and deposited on the shore. I had not seen one like that since childhood, from which it seemed to have buoyed up from some deep memory pool as if this were all a dream.

I am not writing about the Great Flood of 2023 but about a great flood that occurred on the one-year anniversary of that flood, on the night of July 10–11 in 2024. On that day I was sent a photo of an inundated downtown Richmond that I mistook for a photo from the year before, but it had been taken on that morning. The bridge over the Winooski was once again a bridge to nowhere, its trusses rising out of a muddied lake. Over the next days and weeks I passed by the flea market and saw someone tossing stuff into a dumpster, as if I were watching a replay of the events of 2023. This time the damage was even greater, although the Winooski did not rise as high, and it was homes in the uplands that were among the most affected. The water gushed down steep mountain slopes, culverts clogged with silt and debris, or were ripped away, the water taking out homes, roads, and bridges. Acres of topsoil were washed away from fields and pastures, and crops were deluged once again.

My friend Jane lives on the corner where the Huntington River flows into the Winooski, at the bottom of a steep mountain road. Her house was flooded once before—during that Halloween night storm in 2019 when the water gushed down the mountain as it did on this night, but that was only inches of water compared to the four feet she would find in her basement on July 11, 2024. She'd been wide awake, watching the weather, when around midnight she looked out and saw she was surrounded by water. She saw flashing blue-and-red lights down the street, a group of people gathered around, and went out to find out what was going on. When she stepped off her bottom step, she fell into a sinkhole up to her shoulders and injured her foot. She pulled herself out by a post. She remembered how she had felt the house shake earlier, and now she knew her foundation had failed. When she reached the small crowd, she saw a small fire truck had fallen into a ditch.

Her small weather-worn clapboard house looks precarious now, surrounded as it is by new ditches carved out by the floodwater. When I meet with Jane, she sits barefoot on her front step, as if she is afraid to step onto the strange sandpit where her house now sits, and tells me her story. Her foot is a mass of bruises—purple, black, and yellowish blotches from her toes up to her upper ankle. "It looks worse than it is," she says. She smiles as she recounts how a police officer who came to check on her asked her if she had clothes. "He was just going through his checklist," she says. It was our only moment of amusement, but also a reminder of how much worse this climate disaster could have been. Jane must weigh a little over a hundred pounds, and in her bare feet, one of which is a mass of bruises, she also seems frail, but she has endured her ordeal with resilience and equanimity. Her words are practiced by now, as she tells her story as if she were not talking about her own life. She says the support she received from the community and from friends—who showed up to help clear away the muck, who gave her a car because she lost hers, who brought her food, who offered her a place to stay, the four to six weeks she would be displaced turning into four months—was "tremendous."

Jane decided to elevate her house. It is the house where she raised her daughter and where she lived with her husband for more than two decades until he died of cancer in 2011. She is a Russian literature scholar, teacher, and translator and is about to retire. Her house is full of books and papers. Before she and her husband settled here, she moved from place to place, traveling often to the Soviet Union. She was "like a squirrel caching her things in various trees," she tells me. This house is her olive tree, it is the repository of so many memories: The house beside the river she has listened to all these years, a constant whispering. Where she can watch the light on the river through the trees, see it turn to ice and then thaw and swell and diminish and swell again, and where she can watch the kingfishers and the osprey hunt and hear the song of the warblers and the Carolina wren. Where she has watched the point bars—islands of river cobble—shifting and morphing, moving like the hand of a pianist in a slow crescendo until they have grown to monstrous size.

She says people tell her she should leave. But where is she going to go? This is her home. I know that others who have raised their houses have still been flooded as the river moves and the floodwaters reach even higher and I think of this as she speaks. If the past is no longer an indicator of the future, then what do we have to go on? Whatever we do feels risky. The house, lifted up on a new foundation, will at least be aspirational. We will remain, it says. We will endure.

This flood does not even have a name, because names are given to exceptional weather events and we no longer consider floods to be crises, rare events that we suffer through to the other side and return to normalcy. The language we use to talk about storms has changed; flooding is a condition of our lives now, and not only for homes and farms in the floodplains, those high-risk areas. This event was one of multiple bouts of severe flooding in July 2024, and our county was just one of seven counties in Vermont that FEMA then designated as major disaster areas. The photographs in newspapers are starting to look much the same: the muck, the dumpsters, the volunteers in rubber boots. Towns are facing "astronomical" damages that far exceed entire town budgets. "We are turning our pockets inside out," Richmond selectboard member Jay Furr said.

"When you go through multiple 100-year storm events in a couple of years," the assistant secretary for the U.S. Health and Human Services Department said when she visited the ravaged areas, "it is the result of climate change." Because someone needed to say it out loud. No one can feel safe anymore, anywhere, though we still comfort ourselves with assurances. *We are far from the river. We are in a plain and not at the foot of a steep slope.*

We may mark 2023 and 2024 as the years when our paradigms shifted—when we stopped viewing extreme weather events as outliers. While Jane has chosen to stay in place, others in high-risk areas, who have been flooded over and over, have reached their tipping points with this year's floods. Thirteen more homes have been approved by the town of Richmond for FEMA buyouts, with more pending, as well as several who live along the river in Huntington. Other residents want the state to remove gravel bars in the river to reduce the risk of more floods. Rivers move, that is what they do, they move like the hand of a one-handed pianist all over the keyboard, as John McPhee once wrote about the Mississippi. The Huntington is playing a fast and clamorous tune with its one hand, and it's a lot to keep up with. It's a lot to ask human beings to accept this conflation of the human timescale with the geologic. Homeowners have seen their yards washed out, they have seen riparian buffers they planted to stabilize banks torn up and whisked away, like a bear laughing at the efforts of a beekeeper to protect a hive. They are concerned that as the river fills up with gravel and cobbles it is becoming shallower, warmer, and more prone to flooding. The cool clear trout pools are gone. Channels have changed course and in some places the river has reversed direction altogether. The river is moving closer and closer to their front doors. The residents who want the state to help keep the river from moving know that this is only a temporary fix. At some point, they know, they will just have to let the river do what it wants to do.

Flooding is exacerbating the spread of invasives, which are carried by floodwaters that also scour the riverbanks, tilling the soil for them to root and flourish. Knotweed has come to dominate along the Huntington, as it has all over the state since Tropical Storm Irene, accelerating the erosion of river shores. First introduced to help control riverbank erosion, it turns out the weed's weak roots are not good at holding on, and with every flood knotweed is easily ripped away and spread downriver, advancing its forward march—another feedback loop spinning out of our control. And with flooding comes polluted lakes and rivers. The Great Flood of 2023 dumped a year's worth of phosphorus runoff into Lake Champlain in one week. After the 2024 storm dumped acres of topsoil into Lake Champlain—each acre amounting to five hundred dump truck loads—the 121-mile-long "Little Great Lake" turned a thick chocolate brown and did not clear up for weeks. Neither did the Winooski.

One of the houses approved for a FEMA buyout sits on the main road in the village. To see it go this way, I feel like I'm witnessing the removal of a big old tree. For a while its image will still linger as we pass its old place in the village, a phantom house. It does feel like a presaging of a future world without us: all those phantom houses, all across the state, that will year by year fade from our memories, as all memories will do in time.

Song Dogs

When I got home from my walk I picked the last of the poblanos and jalapeños. It is surprising how much food we still have in the garden: we are still picking arugula, kale, chard, bok choy, parsley, white radish, even tomatoes. The nasturtiums spill over the beds where we harvested onions and garlic weeks earlier, flowers that we grow for their beauty but they are also, at this time of year, food—flavorful enough to take the place of tomatoes or cucumbers in a salad. Arugula, white radish, carrots, sunflower seeds, and nasturtium flowers is our standard salad for this time of year.

Our chest freezer is stuffed, loaded with pears and apples, berries, blanched greens—not only kale and chard but also lamb's quarters and dandelion—roasted eggplants, sweet peppers, soups, pesto, sauces, quiches, and pies. This was the best year I can remember for fruit—pears and apples but also raspberries and gooseberries—as if the universe were compensating us for last year's total loss. We have jarred tomatoes and pickles; more pears and apples; gooseberry, raspberry, and wild grape jams; and, stored in our cool basement, fifty-pound bags of onions, baskets full of garlic, and crates of winter squashes. Our chickens are still giving us eggs, and will do so until the long nights of November arrive. The gift of abundance exerts an obligation of labor in return: as Robin Wall Kimmerer reminds us, the earth endows us with gifts, but the other half of that truth is that something is required of us in return. The fruit of the earth will taste sweeter, she says, if we "participate in its transformation." "It is our work, and our gratitude, that distills the sweetness." What we take for granted does not taste as good. "How much better it tastes if you cook it yourself," I remember a student saying in a cooking class at the educational farm next door. And better still if you began with a patch of dirt, some sunlight and water, and a packet of seeds.

After last year's blighted autumn, when the leaves turned brown and dropped, the colors burning like a smoking ash heap rather than a roaring fire, I am especially tuned to the nuances of turning leaves this year. A promising beginning in early fall would disappoint, this time from too *little* rain in September and October, but the softer palette still had a penumbral glow. Autumn should be when the land reaches its climax in splendor, and yet I wish for the season to burn slowly, for the brilliance to last, for daylight savings time not to end. I dread the first hard frost that will kill our dahlias that grow in profusion all around the house. I wish to delay the contraction of my world that comes with winter, when I retire my bicycle. I will miss seeing the youth crews around us, working in the fields or headed for the lean-tos in the woods with their backpacks. I will miss the time I spend in the garden, but I also am grateful for the rest, for my days to be freed for other work. I will also be happy for snow when it comes, and yet I fear that it will not.

The hummingbirds and robins are gone but we are visited by other wildlife at the approach of winter. On one autumn afternoon, I headed for the woods behind the farm next door when I saw a bear at the top of the knoll, standing on her hind legs, reaching for apples. I turned back and headed in a different direction, to leave her in peace, and when I came back an hour later, she was still there, taking a nap under the old apple tree.

Every year it seems there are more and more wild turkeys, who were, a century ago, completely gone, and who tend to stay close to the edge of the woods, where they will run if they are alarmed. But they are not alarmed when I approach them, and I find that I can walk right by them and admire their long, pointed tail feathers that trail behind them like paint brushes.

One night at this time of year we had some friends over for a campfire, and as we stood or sat around the fire, listening to the crackle and pop of the flames, with the perpetual hum of the road in the background, I could hear the distant song of the coyotes. We all paused to listen. The voices were coming from the mountain, and it sounded like a whole chorus of them yelping and howling and yipping. It is at this time of year when the male pups born in spring have reached maturity, and are setting out on their own to mark their own domain, which is when they howl and sing. They are drawing lines on the ground with song. The call-and-response between two coyotes can sound like a whole pack of them—this is an aural illusion called the *beau geste* effect. No wonder the sound is so eerie, so mystical, like the sound of starlight bending.

If it feels like the world contracts in winter, it is true that it also becomes larger and wilder, as the mountains, forests, and rivers reclaim themselves from us. It was lucky we were outside that night when the coyotes chose to sing. There will be other nights, but I won't always be around to hear them. Once I was at a party at a house with a night-blooming cereus, a plant that flowers only one night a year, and this plant decided to bloom on that night, when we were all there to see it. This was a little like that.

The Winooski Abenaki, who were in these woods at this time of year, before the settlers took the land from them, would not have heard this song—it would have been wolves they heard. They would have gathered around fires like this, telling stories, cracking butternuts, and putting their babies to bed on cattail fluff. They, too, gathered the abundance of the season to put away for winter. They hunted bear and deer and small mammals, squirrels, and chipmunks. And gathered fibers and furs and stones for making clothing, baskets, and tools. It is not only the wolves that are gone. The butternuts are rare in these woods now, the beech and the ash trees are dying, and the understory plants never did grow back once the forest was cut. I am anxious about the world. I know how delicate are ecological arrangements. I have seen what an unseasonal frost can do. One set of strings out of tune can wreck a whole symphony. I have seen giant old trees wither and die from beetle infestations fed by a warming world; I know that the sap will stop flowing if the nights are not cold enough; I see how poison parsnip and knotweed are in an unstoppable march to dominate river shores, displacing the cattails that animals need. I fear and mourn the great unraveling. Invasive species are also nature's way of healing, of showing resilience. Another world entirely is rushing toward us, whatever will follow the fossil fuel age, and though it may be beautiful in the end, there will be much suffering as it comes. The coyotes are here because the wolves are gone, but how fierce is their presence. *We are alive*, they are saying to us. *We are here*. It is a kind of *beau geste* effect—that in our domestication, wildness bends over the mountain to come to us as song.

Epilogue
Our Total Eclipse

*A*pril 8, 2024. It was a kind of gestation period. The expectation. The growing reality of it, as the day approached, until you could feel it kicking inside of you. And when it arrived, it was like life had broken through a shell. It was not anything like what we'd been told, over the centuries—darkness, death, plague. The opposite was true.

Earlier in the gestation period, Art and I thought about going out onto the middle of Lake Champlain to experience the total eclipse. We wanted to be under a vast expanse of open sky, and mountaintops are mostly off limits in mud season. But then I thought it might be too terrifying to be floating in a tippy canoe over water hundreds of feet deep. Cold water, with no one else around, as the last patches of snow and ice still clung to the end of winter. I would not be able to grab hold of Art's hand when the moment came.

Eclipses were terrifying events through most of our history, when people did not expect them. The world suddenly, inexplicably, going dark. Even in the modern world, the experience of a total eclipse is said to be terrifying. I got this from Annie Dillard's classic essay on a total eclipse, which I reread only a few weeks before April 8, 2024. She writes that people are known to scream—as they did when she witnessed hers in 1982—when the moon's shadow passes over the world in a *whoosh*, at eighteen hundred miles an hour.

She said, "From all the hills came screams . . . ," then, "The heart screeched . . ." She said:

> We had all died in our boots on the hilltops of Yakima, and were alone in eternity. Empty space stoppered our eyes and mouths; we cared for nothing. . . . Oh, and then the orchard trees withered, the ground froze, the glaciers slid down the valleys and overlapped the towns. If there had ever been people on earth, nobody knew it. The dead had forgotten those they had loved. The dead were parted one from the other and could no longer remember the faces and lands they had loved in the light. They seemed to stand on the darkened hilltops, looking down.

As the day of the eclipse drew near, we made the decision to go to the lake after all. I was no longer afraid, and I wanted a full experience of the event, to be one with the light and the darkness, to be in the cup and grasp of the universe in its spinning.

We pushed off in plenty of time. Only the day before it had been cloudy and cold, and only a few days earlier we had one of the biggest snowstorms of the year. But the clouds parted for us on this day, and the sun came out. Fortunately for us there was no wind; the water as calm as it ever gets. We dipped our paddles and glided over the smooth blue surface of the bay and headed for the open lake, our first day on the water for the season. We were between the Adirondack Mountains on one side and the Green Mountains on the other, with state parks on both shores, and the lake was ours.

I put down my paddle for a moment and put on my eclipse glasses to test them out. The world went completely black—I mean utterly and completely pitch—except for the bright brassy disk high in the sky. My own response was like that of the small child who, when her mother put a pair of eclipse glasses over her eyes and asked her if she saw the sun, said, "No, I see the moon." For that is what it looked like, the moon at night. But oddly.

We were the only boat in sight and this worried me. Did the others know something that we didn't? Would there perhaps be some kind of shudder passing over the world from the shadow rushing over us at eighteen hundred miles an hour? A kind of celestial tsunami, even a gentle one? I knew it was irrational—no one had ever said the eclipse was accompanied by a wind. But there was Annie Dillard whispering in my ear.

As we approached the outlet of Little Otter Creek, we heard voices coming from the shore, where a group of people were gathered on a western-facing beach, a perfect place to be. We paddled over to a small island and rested in its lee shelter. A kayaker and a couple in a canoe appeared from the creek, and I felt some relief that we were not the only ones. I climbed onto the rocks to stretch my legs, and from there I saw the first chip on the bottom right rim of the brassy disk. The chip became a slice, then a bigger slice, as if cut with a razor-sharp knife, the sun diminished into a mere crescent, a scimitar, a scabbard moon hanging in the still bright sky.

It was time, so we pushed off and paddled onto open water, where we would have a full view of the shadow as it ripples over the face of the water. We were chilled by the breeze and regretted we hadn't brought along more clothes. A fishing boat passed us and we exchanged greetings. It seemed that they were fishing, as they might on any ordinary day.

We turned the canoe to face the sun. I looked north, as I expected the shadow would come from that direction. I stared into the light, looking for a change. The platinum film Dillard had described. The moon's shadow kept slicing away at the apple until there was almost nothing left but still it was daylight. It was not even dusk. But what grew upon us was a spectral light, not a twilight but an opaque film over the world that I felt I could brush away with my fingers, a pastel dust. And then the shadow rolled.

"No more glasses," Art said, and I took mine off and looked up at the ring of light, the corona. It was a sleight of hand, a magic trick to exchange a sword for a ring like that, without our noticing the trick. The mountains were silhouetted before an alpine glow, the sky in sunset colors, but strange ones, for this was a sunset on Venus. The darkness that fell over us was not blackness but a multicolored quilt, an opaque watercolor, the lake spread out toward the north, a dusty sapphire in lavender-and-smoke shadow. Behind me the water's surface was embroidered in gold threads, reflected from the delicate nimbus in the sky. I looked up from the pastel drawing I was in to view the ring in its cottony corona. The hole was not black, it was a smoky blue, an aqueous dark, mottled like the surface of the moon. Clinging to the bottom rim was a glowing red teardrop, like the red dot inside an egg. A solar prominence.

And then the lights came back on.

I did not hear any screams. What I heard was a collective gasp, a cheer that rose up from the crowd, like the cheer that erupts when a point guard has just sunk a three-pointer from the logo in the last two seconds of the game. It was the collective cry I have heard on rare occasions in my life, at the sight of something wondrous: the time a group of friends witnessed a monarch butterfly emerge from its chrysalis and fly away; another time when our group, in a small inflatable boat, saw a whale burst through the surface of the water and blow.

And from a distance I heard fireworks. As if they were needed.

I do think that I held my breath. Something was happening. That whale might be passing under us. We could feel a tremor. But it was all too beautiful to feel afraid—the alpenglow, the gold threads on the dark water, the silver nimbus in the sky in its perfection. Three minutes was long enough to have looked toward the mountains, to gaze upward, to turn around and see the ripples on the water outlined in gold ink. I wanted to take it all in so I would remember it, I wanted to be able to turn the light into words. But I could only hold my breath for so long.

And then it was daylight again, just like that—and all that light was coming from just a sliver of the sun. You could not tell that the sun was 90 or 95 percent eclipsed. Without your eclipse glasses, this was just an ordinary day.

At the end of Dillard's essay, she writes that as soon as the eclipse passed totality, and the shadow began its retreat, everyone just turned back and headed home. They got in their cars. They headed on down the mountain. "The sun was still partially eclipsed—a sight rare enough, and one which, in itself, we would probably have driven over five hours to see." But it was all over for them. And this is what we did. As soon as the lights came back on, I said to Art, "Let's go back." Or was it Art who said that to me? We were cold, very cold, and now what we wanted to do was to paddle hard to get warmed up, to get out of the canoe we had been sitting in for more than two hours. We would still look up through our glasses to view the retrograde movement of the moon's shadow, but it was over.

I felt I had been lifted—transported out of time, away from the earth and all its troubles for those three minutes. Later, when I listened to the reports on the radio from all over Vermont, I was glad to have felt myself in the company of so many others who experienced the same rhapsody, from their hilltops or beaches or parking lots or balconies. I was thankful for the astronomers who told us what to expect and when, with such precision, so that we were prepared and we were not afraid. The dead did not look down on us. There were no conspiracy theories. No one would say the eclipse did not happen. I was glad for Vermont, that our happiness was not ruined this time by the weather and we were not disappointed. We all got to be astronauts for a day.

I understood why, when the lens cap began to slide backward, everyone turned back. "Enough is enough," Dillard writes. Even of glory. We had watched the progress of the eclipse with such sustained concentration that now we needed to let our minds go slack. The climax of the eclipse—*totality* (this was the new word in our vocabulary)—was so enormous that we were empty after that. Ninety percent was not worth our attention. "From the depths of mystery, and even from the heights of splendor, we bounce back and hurry for the latitudes of home." I too was happy to be back on terra firma, in the familiar world, in the latitudes of the earthlings.

The world was still here, but I had changed, because when you see something wondrous, you are not the same person that you were. And those latitudes, the home that you had been given or that found you after a lifetime of looking, had prepared you for this, and for this you loved it more than you ever had before.

NOTES

PREFACE

xiii   The anthropologist David Graeber calls this "play farming": David Graeber and David Wengrow, *The Dawn of Everything: A New History of Humanity* (Farrar, Straus and Giroux, 2021), 260.

xiv   The ideas expressed by E. F. Schumacher: Schumacher, E. F. *Small Is Beautiful: Economics as if People Mattered*, introduction by Theodore Roszak (Harper, 1973), 4.

COMING HOME

6   Some historians would say that "thinkers": Gary Snyder, *The Practice of the Wild* (Farrar, Straus and Giroux, 1990), 61.

9   remind us that this is what farmers: Sharon Aysk and Aaron Newton, *A Nation of Farmers: Defeating the Food Crisis* (New Society Publishers, 2009), 10–11.

10   "The ecology of freedom": Graeber and Wengrow, *Dawn of Everything*, 260.

10   *You can't do just one thing*: Michael Pollan, *The Omnivore's Dilemma: A Natural History of Four Meals* (Penguin, 2006), 225.

10   Gifts occur in "a realm of humility and mystery": Robin Wall Kimmerer, *Braiding Sweetgrass* (Milkweed, 2013), 24.

11   a bowl of beans: Kristen Kimball, *The Dirty Life: A Memoir of Food, Farming, and Love* (Scribner, 2010), 269.

11   "As our domestic animals settled in": Carl Safina, *Beyond Words: How Animals Think and Feel* (Picador, 2015), 235.

12   George Orwell, near the end of his shortened life: Rebecca Solnit, *Orwell's Roses* (Viking, 2021), 247.

12   He would teach generations of writers: "Politics and the English Language," in *A Collection of Essays by George Orwell* (Harcourt Brace, 1946).

12   "Our job," Orwell wrote in his essay on Gandhi: George Orwell, "Reflections on Gandhi," in *A Collection of Essays*, 175.

OFF THE TRAIL

18   We are free to find our own way: Gary Snyder, "Off the Trail," in *No Nature* (Penguin Random House, 1992), 369–70.

A BARN, A NEST, A MOORING

21   "The old barn is fast disappearing": Thomas Durant Visser, *Field Guide to New England Barns and Farm Buildings* (University Press of New England, 1997), 58.

25   *The American Barn,* also has photos: Randy Leffingwell, *The American Barn* (Motorbooks International, 1997), 178–79.

25   until they owned "nearly the entire valley": Leffingwell, *American Barn,* 176–78.

27   As one Connecticut farmer wrote: Visser, *Field Guide,* 7.

29   "And so when we examine a nest": Gaston Bachelard, *The Poetics of Space,* trans. Maria Jolas (Penguin, 2014), 123.

29   "the nest is not built until later": Bachelard, *Poetics,* 114.

MESCLUN DAYS

37   "occurred to me while picking beans, the secret of happiness": Kimmerer, *Braiding Sweetgrass,* 121.

39   "A bowl of beans, rest for tired bones": Kimball, *Dirty Life,* 269.

39   Robin Wall Kimmerer says the reason: Kimmerer, *Braiding Sweetgrass,* 122–24.

43   In his poem "Spinazzola: Quella Cantina La": Richard Hugo "Spinazzola: Quella Cantina La," in *Selected Poems* (W.W. Norton & Co., 1979), 63.

43   Italy in 1963 was a place "full of fountains": Richard Hugo, *The Triggering Town: Lectures and Essays on Poetry and Writing* (W.W. Norton & Co., 1979), 76.

44   I came upon an essay almost twenty years ago: James Lasdun, "Diary: Salad Days," *London Review of Books,* February 9, 2006.

SHEARING DAY

50   An old-timer interviewed in the Foxfire books: Eliot Wigginton, ed., *Foxfire 2* (Anchor, 1970), 179.

FOOD AS MEDICINE

59   "Such a solitary mind—if it could exist . . .": Gary Snyder, *Practice,* 60.

NOTES

ASTERIX

65   the author's encounter, while walking alone: Naomi Mattis, "Moonfire," in *Intimate Nature: The Bond between Women and Animals*, ed. Linda Hogan, Deena Metzger, and Brenda Peterson (Ballentine, 1998), 280–84.

THE MONITOR ELM

73   In his beautiful essay: Loren Eiseley, "The Brown Wasps," in *The Best American Essays of the Century*, ed. Joyce Carol Oates (Houghton, 2000), 239–45.

74   "It was part of my orientation": Eiseley, "Brown Wasps," 245.

74   Gachelon Bachelard writes, "The suffering tree": quoted in Michael Perlman, *The Power of Trees: The Reforesting of the Soul* (Spring Publication, 1994), 126.

74   "before Jesus, before Rome": William Stafford, "B.C." in *Stories That Could Be True: New and Selected Poems* (HarperCollins, 1977), 76.

UNDER THE PENOBAGOS

76   under the Penobagos, the Moon: Frederick M. Wiseman, *The Voice of the Dawn: An Autohistory of the Abenaki Nation* (University Press of New England, 2001), 93.

76   In my own time archeologists found: Peter A. Thomas, "Richmond's Ancient Past," in *Richmond, Vermont: A History of More than 200 Years*, ed. Harriet Riggs (Richmond Historical Society, 2007), 4–10.

77   gardening is a way of loving the land: Kimmerer, *Braiding Sweetgrass*, 123.

78   when you say "I own this": Margaret Atwood, "The Moment," https://poetryarchive.org/poem/moment/.

78   Reciprocity, respect, reverence: Judy Dow, "Going through the Narrows," *Potash Hill* (Spring 2019), https://potash.emerson.edu/2019/spring/narrows/.

79   they remember the Great White Bear: Wiseman, *Voice of the Dawn*, 16.

80   Mobility, they knew, was the key to conservation: William Cronon, *Changes in the Land: Indians, Colonists, and the Ecology of New England* (Hill and Wang, 1983), 37–38.

LOOKING AT LAMBS

89   In his classic essay: John Berger, *Why Look at Animals?* (Penguin Great Ideas, 2009), 12–37.

90   "Not assuming they have thoughts": Safina, *Beyond Words*, 27.

90   Species differ, Safina writes: Safina, *Beyond Words*, 324.

91   "Wolves and humans can understand": Safina, *Beyond Words*, 237.

92   to borrow from Joel Salatin: Pollan, *Omnivore's Dilemma*, 218–19. Michael
     Pollan, "An Animal's Place," *New York Times*, November 10, 2002.

93   "Once the animals flowed": John Berger, "They Are the Last," in *Why Look
     at Animals?* 79.

                            GOING GENTLY

95   "When I got home from the airport": Alexis Lathem, "Coming Home,"
     Alphabet of Bones (Wind Ridge, 2015), 50.

97   "Such a round of chores is not": Snyder, *Practice*, 153.

99   The heart of the world lies open: Charles Wright, "The Evening Is Tranquil,
     and Dawn Is a Thousand Miles Away," *Sestet*, 25.

99   The last scud of day holds back: Walt Whitman, "Song of Myself," *Leaves of
     Grass* (final "Death-Bed" edition, 1891–1892) (David McKay, 1892), https://
     www.poetryfoundation.org/poems/45477/song-of-myself-1892-version.

                            GRASS FARMING

105 "Suddenly, as he worked": Tolstoy, Leo, *Anna Karenina*, trans. Joel Carmi-
     chael (Bantam, 1960), 267.

107 "A good solution solves more than one problem": Wendell Berry, "Solving
     for Pattern," in *The Gift of Good Land: Further Essays Cultural and Agricultural*
     (North Point Press, 1981), 5.

107 who once "flowed like their milk": John Berger, "They Are the Last," in *Why
     Look at Animals?* 79.

108 "If the sixteen million acres now": Pollan, *Omnivore's Dilemma*, 197–98.

108 once the soils are healthy, they are merely storing it: Tara Garnett et al.,
     Food Climate Research Network, "Grazed and Confused?" (2017), 32, https://
     www.oxfordmartin.ox.ac.uk/downloads/reports/fcrn_gnc_report.pdf.

108 the world's 1.5 billion cattle alone: "Agriculture and Aquaculture: Food
     for Thought," EPA, October 2020, https://www.epa.gov/snep/agriculture
     -and-aquaculture-food-thought#.

108 the "two percent solution": Courtney White, *Grass, Soil, Hope* (Chelsea
     Green, 2014), 7, Marine Carbon Project, https://marincarbonproject.org/.

109 Another is TomKat Ranch: TomKat Ranch, https://www.tomkatranch
     .org/.

109 "Imagine what it would mean if a net-zero-emissions": quoted in Michael
     Grunwald, "Tom Steyer Thinks His Ranch Can Save the Planet," *Politico*,
     October 11, 2019.

109 as Politico reported in a 2019 article: Grunwald, "Tom Steyer," *Politico*.

109 John Wick, too, would have to temper his claims: Charlie Siler, "The
     Marin Carbon Project's Mixed Prognosis for the Future of Agriculture,"

Craftsmanship (2024), https://craftsmanship.net/sidebar/the-marin-carbon
-projects-mixed-prognosis-for-the-future-of-agriculture/.

109 **MIT Climate portal concludes**: MIT Climate Portal, "Soil-Based Carbon Sequestration," April 15, 2021, https://climate.mit.edu/explainers/soil-based-carbon-sequestration#.

110 **A comprehensive 127-page global study**: Garnett, "Grazed and Confused?."

110 **Another 2023 study that compared emissions**: Pardo, Guillermo et al. "Carbon footprint of transhumant sheep farms: accounting for natural baseline emissions in Mediterranean systems," *International Journal of Life Cycle Assessment*, 2023.

110 **cows, sheep, hogs, and poultry are not equal partners**: Isabelle Gerretsen, "What Is the Lowest Carbon Protein?" BBC December 15, 2022, https://www.bbc.com/future/article/20221214-what-is-the-lowest-carbon-protein.

111 **"The revolution won't happen by people staying home"**: Rebecca Solnit, "Big Oil Coined 'Carbon Footprints' to Blame Us for Their Greed," *Guardian*, August 23, 2021.

111 **This is the meaning of *direct action***: David Graeber, *Direct Action: An Ethnography* (AK Press, 2009), 202.

CONSIDER THE LAMBS

112 **in my midtwenties, after reading**: Frances Moore Lappé, *Diet for a Small Planet* (Ballantine, 1971).

113 **I will never get over reading**: Melissa Coleman, *This Life Is in Your Hands* (Harper, 2011), 110.

113 **Even Peter Singer, the ethicist who wrote**: quoted in Pollan, *Omnivore's Dilemma*, 327.

114 **"Don't let anyone tell you that growing vegetables"**: Kimball, *Dirty Life*, 170.

115 **"For years I'd taken imagery on dairy-milk cartons literally"**: Lisa Monroy, "Soli/dairy/ty," *Long Reads*, February 24, 2020, https://longreads.com/2020/02/24/soli-dairy-ty/.

115 **"The happy calf sucking its dam"**: Brad Kessler, *Goat Song: A Seasonal Life, A Short History of Herding, and the Art of Making Cheese* (Scribner, 2009), 50.

116 **"Mark wrapped a towel around the heifer"**: Kimball, *Dirty Life*, 180.

117 **In the Roquefort region**: Mark Kurlansky, *Milk! A 10,000 Year Food Fracas* (Bloomsbury, 2019), 285–86.

118 **"I wonder if, by attempting to raise two vegans"**: Monroy, "Soli/dairy/ty."

118 **"From the cow comes a whole farmstead of abundance"**: Kimball, *Dirty Life*, 91–92.

119 **"You know, o my son, the first thing God created"**: quoted in Jim Corbett, *Goatwalking: A Guide to Wildland Living* (Viking, 1991), 70.

119 **"Archeologists generally maintain that"**: Corbett, *Goatwalking*, 8.

119 **Humans adapted agriculture slowly**: Graeber and Wengrow, *Dawn of Everything*, 210–48.

119 **domestication was not an "invention"**: Stephen Budiansky, *The Covenant of the Wild: Why Animals Chose Domestication* (Yale University Press, 1999), 43–68.

120 **"A handful of minor species emerged"**: Budiansky, *Covenant of the Wild*, 61.

120 **"Carnivorism," the Nearings wrote**: Helen and Scott Nearing, *The Good Life: Helen and Scott Nearing's Nearly Sixty Years of Self-Sufficient Living* (Shocken, 1989), 142.

121 **John Berger calls this paradox**: Berger, *Why Look at Animals?*, 16.

121 **"Pastoral metaphors of the nomad religions"**: Corbett, *Goatwalking*, 43.

121 **What is it about herding, Kessler asks**: Kessler, *Goat Song*, 170.

122 **"That's it. That's the pinch in the hourglass"**: quoted in Dorothy Wickenden, "Wendell Berry's Advice for a Cataclysmic Age," *New Yorker*, February 21, 2022.

123 **The World Resource Institute estimates**: Grunwald, *Politico*.

123 **an annual 2,700 tons of unrecyclable plastic**: Annie Macmillan, "Vermont Act 731: Creation of an Agricultural Plastics Collection and Recycling Program," PowerPoint presentation, February 4, 2020, 15, https://tpsalliance. org/wp-content/uploads/2020/02/Track-A-Session-5a-Macmillan.pdf.

123 **The farmed landscape of the Lake District**: James Rebanks, *A Shepherd's Life: Modern Dispatches from an Ancient Landscape* (Flatiron, 2015).

124 **James Rebanks, a shepherd of the district**: Rebanks, *A Shepherd's Life*, 15.

LA DIGNIDAD

130 **As Vermont Governor Peter Shumlin would say**: "Vt. Gov. Wants Police Directive on Migrant Workers," *NBC News*, September 16, 2011.

132 **From 2020 to 2023 Vermont lost 128 dairy farms**: Emma Cotton, "Lost My Love for Farming," *VTDigger*, February 9, 2023, https://vtdigger.org /2023/02/09/lost-my-love-for-farming-the-fate-of-dozens-of-small-org anic-dairy-farms-may-rest-with-vermonts-budget-adjustment-act/.

132 **More than half of the 2.5 million seasonal workers**: Immigration Reform and Farmworkers (2024), Farmworker Justice, https://www.farmworkerjustice. org/advocacy_program/immigration-reform-farmworkers/.

132 **"One of the goals of immigration restrictions"**: Migrant Justice, "We Remain Here Today, and We Will Continue Fighting to Remain," November 17, 2024, https://migrantjustice.net/news/%E2%80%9Cwe-remain-here-today-and-we -will-continue-fighting-to-remain.

133 **"This is the first year that I can take a vacation"**: Migrant Justice, "Testimonials," n.d., https://migrantjustice.net/testimonials.

134 **"American agriculture's like a big pie"**: quoted in Robert Wolf, "Farming as a 'Scientific Business,'" *North American Review* 285, no. 2 (March–April, 2000), 45.

135 **The architects of NAFTA on the Mexican side**: Alyshia Galvez, *Eating NAFTA: Trade, Food Policies, and the Destruction of Mexico* (University of California Press, 2018), 13.

135 **Over the next decade and a half**: Public Citizen, "NAFTA's Legacy," https://www.citizen.org/wp-content/uploads/NAFTA-Factsheet_Mexico-Legacy_Oct-2019.pdf.

135 **The destruction of the traditional milpa- based food system**: Galvez, *Eating NAFTA*, 2ff.

136 **"Huertas allows the possibility"**: Teresa M. Mares, *Life on the Other Border: Farmworkers and Food Justice in Vermont* (University of California Press, 2019), 96.

136 **In his 2018 book *Down to Earth***: Bruno Latour, *Down to Earth: Politics in the New Climactic Regime* (Polity, 2018), sect. 2.

137 **Mexico is expected to lose half of its farms**: "How Is Climate Change Affecting Mexico?," Climate Reality Project, February 15, 2018, https://www.climaterealityproject.org/blog/how-climate-change-affecting-mexico#.

MONARCHS AND MILKWEED

139 **It was Toronto-based zoologist**: Fred A. Urquhart, "Monarch Butterflies Found at Last," *National Geographic*, August, 1976, 160–73. Monarch_Butterflies_Found_at_Last_the_Monarchs_Winter_Home_-_article.pdf.

140 **"I gazed in amazement at the sight"**: Urquhart, "Monarch Butterflies Found at Last," 2.

141 **"Few things indeed have I known"**: Vladimir Nabokov, "Butterflies: The Childhood of a Lepidopterist," *New Yorker*, June 5, 1948.

141 **"A lightbulb moment for me"**: Veronica Kyle and Laurel Kearns, "The Bitter and the Sweet of Nature: Weaving a Tapestry of Migration Stories," in *Grassroots to Global: Broader Impacts of Civic Ecology*, ed. Marianne E. Krasny (Comstock, 2018), 49.

141 **For the Mexican American artist**: Olivia Barrionuevo, *Monarch*, n.d., https://www.oliviabarrionuevo.com/projects_category/1-the-monarch/.

141 **"an example of how fully connected life is"**: Barrionuevo, *Monarch*.

142 **"Every wide-eyed child and meadow walker"**: Urquhart, "Monarch Butterflies," 1.

142 **researchers have identified the break**: Anurag Agrawal, *Monarch and Milkweed* (Princeton University Press, 2017), 227–29.

A CAUTIONARY TALE

146   **two hundred years ago this was all sheep pasture:** Robert F. Balivet, "The Vermont Sheep Industry: 1811–1880," *Vermont History: Proceedings of the Vermont Historical Society* 33, no. 1 (January, 1965): 243–49.

146   **a mass of rock as large as the Great Pyramids:** Tom Wessels, *Reading the Forested Landscape: A Natural History of New England* (Countryman, 1997), 59.

146   **But "pastoralism," he reminds us:** Corbett, *Goatwalking*, 34.

147   **William Jarvis, the Spaniard who:** Balivet, "Vermont Sheep," 243.

148   **As I walk these interconnected trails:** Colin G. Calloway, *The Western Abenakis of Vermont, 1600–1800: War, Migration, and the Survival of an Indian People* (University of Oklahoma Press, 1990), 19.

148   **The Clovis people, who migrated from Asia:** Tim Flannery, *Eternal Frontier: An Ecological History of North America* (Atlantic Monthly, 2001), 183–84.

148   **though new evidence suggests:** Cecily Hilleary, "Native Americans Call for Rethink of Bering Strait Theory," *VOA News*, June 19, 2017, https://www.voanews.com/a/native-americans-call-for-rethink-of-bering-strait-theory/3901792.html.

148   **The sacred ecology of Indigenous peoples:** Fikret Berkes, *Sacred Ecology* (Routledge, 2021), 242–46.

149   **The New England farmers who slashed:** Jane Brox, *Clearing Land: Legacies of the American Farm* (Farrar, Straus and Giroux, 2004), 40–41.

149   **the *code of the honorable harvest*:** Kimmerer, *Braiding Sweetgrass*, 187.

TREEFALL

155   **"The more you study delight":** Ross Gay, *Book of Delights* (Coronet, 2020), xii.

SPRING SURPRISE

160   **But cultures that relied on current sunlight:** Jeremey Rifkin, *Entropy: Into the Greenhouse World* (Bantam, 1980), 113.

IN THE TIME OF CORONA

167   **A forest, writes David Quammen:** Nicolas Triolo and David Quammen, "Why David Quammen Is Not Surprised," *Orion*, March 17, 2020, https://orionmagazine.org/article/why-david-quammen-is-not-surprised/.

168 **"I will probably take my children"**: quoted in Charlotte Alter, "How Conspiracy Theories Are Shaping the 2020 Election—and Shaking the Foundation of American Democracy," *Time*, September 10, 2020.

168 **bitcoin alone has consumed**: Jon Huang, Claire O'Neill and Hiroko Tabuchi, "Bitcoin Uses More Electricity Than Many Countries," *New York Times*, September 3, 2021; Ryan Browne, "Why Big Tech Is Turning to Nuclear Energy to Power Its Energy-Intensive AI Ambitions," *CNBC*, October 16, 2024, https://www.cnbc.com/2024/10/15/big-tech-turns-to-nuclear-energy-to-fuel-power-intensive-ai-ambitions.html.

169 **during the time of the Underground Railroad**: Angela Evancie, "What's the History of the Underground Railroad in Vermont?" *Brave Little State*, *Vermont Public*, October 6, 2017.

170 **In the event of a bird flu outbreak**: Jason Mark, "Egg Prices Are Soaring. Are Backyard Chickens the Answer?" *Civil Eats*, February 18, 2025, https://civileats.com/2025/02/18/op-ed-egg-prices-are-soaring-bring-out-the-backyard-hens/.

THE GREAT FLOOD, AND OTHER CATASTROPHES, 2023

180 **"In my 25 years of working with fruit crops"**: Agency of Agriculture, Food and Markets, "Vermont Crop Damage Could Be Far-reaching after Mid-May Frost," May 23, 2023, https://agriculture.vermont.gov/agency-agriculture-food-markets-news/vermont-crop-damage-could-be-far-reaching-after-mid-may-frost#.

180 **exceeded all fire records by a "stupendous" margin**: Brent McDonald, Matt Joycey, and Brett Laffin, "Canada Is Ravaged by Fire. No One Has Paid More Dearly Than Indigenous People," *New York Times*, July 29, 2023.

181 **Look at a chart comparing 2023**: Benjamin Shingler and Graeme Bruce, "Five Charts to Help Understand Canada's Record-Breaking Wildfire Season," *CBC News*, October 19, 2023.

181 **"Never before has a snowstorm smelled"**: Ahmad Mukhtar, "Canada Wildfires Never Stopped, They Just Went Underground as 'Zombie' Fires Smolder on through Winter," *CBC News*, February 23, 2023.

181 **Climate scientists think**: Oliver Milman, "After a Record Year of Wildfires, Will Canada Ever Be the Same?" *Guardian*, November 8, 2023. Brendan Byrne, et al., "Carbon Emissions from the 2023 Canada Wildfires," *Nature* 633 (2024): 835–39.

182 **But here in the Northeast the climate**: Fourth National Climate Assessment, chapter 18, "Northeast," https://nca2018.globalchange.gov/chapter/18/.

MOVING RIVERS, PHANTOM HOUSES

191 **"When you go through multiple 100-year"**: quoted in Auditi Guha, "From St. Johnsbury to Lyndonville, State and Federal Officials Visit Flood-Ravaged Communities to Discuss Health and Mental Health Challenges," *VTDigger*, August 6, 2024, https://vtdigger.org/2024/08/06/from-st-johnsbury-to-lyndonville-state-and-federal-officials-visit-flood-ravaged-communities-to-discuss-health-and-mental-health-challenges/.

191 **Other residents want the state**: Diane Reynolds, "Fed Up with Flooding," *Times Ink*, October, 2024.

191 **they move like the hand**: John McPhee, *The Control of Nature* (Farrar, Straus and Giroux, 1989), 5.

192 **the spread of invasives**: Josh Crane, "A Rogue Gallery: Vermont's Most Destructive Invasive Species," Brave Little State, *Vermont Public*, October 11, 2024, https://www.vermontpublic.org/podcast/brave-little-state/2024-10-11/a-rogues-gallery-vermonts-most-destructive-invasive-species.

192 **The Great Flood of 2023 dumped a year's worth**: Abigail Giles, "Recent Flooding Will Worsen Lake Champlain Water Quality, but Not as Bad as Last Year," *Vermont Public*, July 24, 2024.

SONG DOGS

193 **if we "participate in its transformation"**: Kimmerer, *Braiding Sweetgrass*, 69.

EPILOGUE

198 **"From all the hills came screams"**: Annie Dillard, "Total Eclipse," in *The Best American Essays of the Century*, ed. Joyce Carol Oates (Houghton, 2000), 482.

201 **"The sun was still partially eclipsed"**: Dillard, "Total Eclipse," 489.

201 **"From the depths of mystery"**: Dillard, "Total Eclipse," 489.

ACKNOWLEDGMENTS

I am grateful to my editor at the University of Massachusetts Press, Matt Becker, for his faith in this project from my initial query. He has given me encouragement and guidance throughout the process of making and publishing this book. I am grateful to Ivo Fravashi for her careful editing, and to the entire production team for their contribution to the making of this book.

A fellowship from the National Endowment for the Arts allowed me to tip the balance between time spent on bread labor and on creative pursuits, in favor of writing. For this unexpected gift I am deeply grateful.

Thank you to my sister Laurie, who read drafts of many of these essays, for her always reliable feedback. To Teresa Mares at the University of Vermont for reading "La Dignidad." To Ian Stokes for reviewing my pages on trail ethics and soil carbon sequestration. And to Brian Tokar for reading chapters on Vermont's ecological history and climate change. Thank you to the peer reviewers who made many valuable suggestions for revision.

I am grateful to my friend and CCV faculty colleague Jane Miller for allowing me to share her story. And to the people of Richmond whose flood traumas are touched on in these pages. I hope these words will expand the circles of empathy for everyone directly affected by climate change.

To the staff, crew members, and volunteers at the farm at VYCC who gave me their time over the week I immersed myself with them, especially Will Lintilhac, Olivia Bulgur, Jeremy Schleining, Nicole Mitchell, Cae Keenan, and Paul Feenan. Also to Jake Kornfeld and Breck Knauft. When Art first moved here, he expected he would be one day surrounded by subdivisions. Instead, our spirits are lifted every time we look out to see youth crews at work and a land restored. We could not ask for better neighbors. Thank you to the Vermont Department of Historic Preservation for supporting our little barn restoration, and to the Richmond Land Trust for saving the monitor barn, with the support of Senator Jim Jeffords, who had a vision of what an architectural beauty like that could do for a whole community.

Even a very small farm such as ours requires a community, and we could not do this without the help of our neighbors (we do have good neighbors without good fences) and our tenants, who look after our animals while we're gone. John and Amy Dodson, Ben Parsons, Timothy Parsons, Aleah Papes, and Conor Emerson are the ones who have lived here the longest and deserve special mention. Also Cody Quattrocci, who helped us to recover our wayward Soay lamb on that day. Thank you to Gerry Guillemette for supplying our flock's winter hay all these years. And to the volunteers at Richmond Fire and Rescue who saved one of our ewes who had fallen into a hole. It took them three hours.

It was my city friends who first nudged me to write stories about our animals, and whose enjoyment of my early pieces about lambs encouraged me to keep writing about our farm. Thank you to my friends Gioia Kuss and Don Kenny, who taught me most of what I know about gardening and who have been there for me since the first Vermont chapter of my life without fail. I am grateful to the Black Earth Institute community of writers for their dialogue, support, and inspiration. And to Judy Dow for her teachings on land acknowledgment. For giving us her winning raffle ticket, thank you to Hope Cifro.

As I complete a book that looks back over many years, I am reminded of the many people who have been important to the prequel of my Ewetopia story and to the making of this book. I want to thank Ellen Taggart, who took a chance on me when she hired me as staff writer at Rural Vermont years ago. And earlier still, I am grateful to Anne Petermann, who, when I showed up to volunteer at the Native Forest Network in 1993, expecting to be given a job stuffing envelopes, asked me to write an article on Hydro Quebec, my first foray into journalism. And to my college friend Hollis Hope and her partner John Davis, who invited me to visit them in Vermont and then to stay for one whole blissful summer. I wrote my first poems on their back deck overlooking the Adirondack Mountains. To my sister Fré, who made it possible for me to complete my MFA, I am deeply grateful. To Guy Jones, who did indeed introduce this New Yorker to mesclun, and to whom I owe thanks for my own epiphany while growing beans.

To the Richmond Racial Equity Committee: our work will be needed more than ever in the coming years. And to the Richmond Climate Action Committee, who have helped me to appreciate what it will take to rebuild a world built on fossil fuels, one sidewalk, one zoning change, and one leaky old building at a time. I am also blessed to live in a community with an excellent public library, where I found many of the books that informed the writing of this one.

Most of all, I am grateful to my husband Art. He has lived with this book, both on and off the page, for as long as I have. This is a life we have built together. Thank you for making it all possible. This book is for you.

ALEXIS LATHEM is an essayist, poet, journalist, teacher, activist, gardener, and craftsperson. She is the author of the poetry collection *Alphabet of Bones* and two chapbooks, and holds an MFA in Writing from Vermont College. Her work has been supported by the National Endowment for the Arts, the Vermont Arts Council, the Black Earth Institute, the Bread Loaf Writer's Conference, the Marble House Project, and the Chelsea Award for Poetry. She has worked in food and environmental justice organizations for many years and reported widely on farm policy and environmental conflicts in Indigenous communities. Her essays and poems have appeared in *About Place*, *AWP Chronicle*, *The Hopper*, *Hunger Mountain*, *Gettysburg Review*, *Solstice*, *West Branch*, and elsewhere. She lives with her musician husband on a small farm in the Winooski River Valley, ancestral land of the Abenaki.